Building the Apollo Capsules

Building the Apollo Capsules

An Engineer's Memoir of the Moonshot Program and Its Debt to Hispanic Team Members

Jim De La Rosa

McFarland & Company, Inc., Publishers
Jefferson, North Carolina

ISBN (print) 978-1-4766-8719-3
ISBN (ebook) 978-1-4766-4871-2

Library of Congress and British Library
cataloguing data are available

Library of Congress Control Number 2023020341

Front cover: The Apollo 17 Command and Service Module (CSM) orbiting the Moon on December 15, 1972; (background) diagram of the Apollo CSM (NASA)

Printed in the United States of America

McFarland & Company, Inc., Publishers
Box 611, Jefferson, North Carolina 28640
www.mcfarlandpub.com

In memory of my late loving wife,
Maria Luisa Figueroa De La Rosa

Table of Contents

Acknowledgments

In early 2013, I stopped writing what I thought was one book that turned out to be three books. I put aside what was to be the third book and printed copies of the manuscript that I had treated as a single book at the time so that I could have people read it and give me some comments. I chose to start with family members, so in March 2013 I gave a copy of the manuscript to two relatives to get their feedback. Then, in August 2013, I gave a copy of the manuscript to two other family members. In November 2014, I gave a copy of my manuscript to a fifth member of my family for the same purpose, and many months later two relatives returned the manuscript with their comments, which I incorporated into the text. One was my youngest brother, Rudy De La Rosa, and the second one was Jimmy Downing, who is married to one of my nieces. To Rudy and Jimmy, I would like to extend a "thank you" from the bottom of my heart for their effort because it was a large manuscript and took many hours to read.

Because of personal problems and other responsibilities, I could not devote much time to my book, and it was not until 2017 that I could try in earnest to find a publisher. As normally happens to first-time authors, I received many rejections. However, on July 12, 2018, one publisher, Cynren Press, made some comments about the manuscript that were meaningful. Some of the comments were that the manuscript would require much developmental editing to get it in shape to publish and that it would be outside of their time allowance. They also wrote that the manuscript was too personal for the subject. This was the first publisher to offer feedback on my manuscript, so I kept their comments in mind. The second publisher to comment on my manuscript in 2019 was McFarland & Company in North Carolina. The editor I had been communicating with was Dré Person, and he told me that McFarland could not publish my manuscript because the editors felt that I had two books in my manuscript—one book was based on the design and flights of the first American cruise missile and a second book was on the Apollo capsules. Dré told me that McFarland would be interested in either of the two books and asked me to determine which one I would like to separate out of the manuscript and present to McFarland. I chose the Apollo capsules book and got back to Dré, and we agreed on a plan. Dré guided me through months of editing and revising my Apollo capsules manuscript—even to the point of making it less personal, as a previous publisher had also suggested—until the manuscript was ready for publication. I want to thank Dré Person personally for sticking with me and all his wonderful help because I know that without him, I would not be a published author. I also would like to give my thanks to Cynren Press for their two comments about my manuscript, because they raised a red flag for me, and as soon as Dré Person of McFarland made his comments to me, I knew exactly what I had to do to get my manuscript published.

Most of the photographs that are part of this book belong to the Boeing Company, and in 2018 I made my first attempt to get approval for some Boeing photographs. To do that, I first contacted the legal department of the Boeing Space Division, which is in Seal Beach, California. I communicated with a person named Lisa Colby, who told me she would take care of my request. Several days later, I received a phone call from Michael Lombardi from the Boeing Space Museum in Seattle, Washington, who told me that Lisa had asked him to call me regarding my request. Michael asked me to email him the photographs so he could review them for me. On February 5, 2018, I received an email from Lisa stating that she and Michael approved the use of the Boeing photographs in my book. Thereafter, whenever I had Boeing photographs to be approved, I would send them to Michael, and he would have Heather Anderson of Boeing Intellectual Property Copyright and Trademark Licensing review and approve the images. My heartfelt thanks go to Lisa Colby, Michael Lombardi, and Heather Anderson for their much-appreciated help in giving me the approval for the use of all the Boeing photographs in this book.

Finally, I would like to thank Jonesy Worrall of the Honeywell Aerospace legal department in Minneapolis for allowing me to use three Honeywell photographs.

Preface

As the world population sat awestruck in front of their TV sets on July 16, 1969, and watched the launch of Apollo 11 and listened to the thunderous roar of the Saturn V booster engines as they propelled Neil Armstrong, Edwin Aldrin, and Michael Collins on humanity's most adventurous and significant journey to the moon, four engineers from North American Rockwell from the Guidance and Control Department from the Apollo program—Paul Garcia, Mert Stiles, George Cortes and myself (Jim De La Rosa)—were in a lab in one of the buildings of the North American Rockwell Space Division in Downey, California. We were there with many other company engineers observing the launch and mission of Apollo 11 on TV monitors that had been placed in the lab for that purpose. We four engineers were there to monitor the operation of the Apollo Spacecraft Guidance and Control System so we could help the engineers at the NASA Manned Spacecraft Center in Houston, Texas, if they needed our technical assistance in the event of a system failure that could jeopardize the mission or the lives of the crew. Our company had designed the Apollo capsules, the Service Module, and the Saturn II booster—three key parts of the rocket that were part of the Apollo 11 mission.

At that moment, during the launch of Apollo 11, I was wondering deep in my heart how many Hispanics were watching. Probably millions were watching worldwide, but how many of them realized, at that moment of the launch, the significance of the event and the part that Hispanics played in it? The event was dramatic, one of America's greatest achievements of the 20th century, real and now part of history. It was the culmination of years of effort and a tremendous amount of behind-the-scenes work by technicians, engineers, managers, executives, and the people who manufactured the moon-bound machines. Without the effort of designing, managing, manufacturing. and preparing a spaceship for a trip to the moon and returning safely to Earth, the lunar expedition would not have been possible. It is in the behind-the-scenes work that Hispanics contributed and excelled. However, this fact has never been publicly documented, and my concern is that by the effective method of exclusion, we Hispanics are again in jeopardy of losing our due recognition. Hispanics were not just witnessing history; we were also making history by helping to put humans on the moon in 1969.

This book tells the story of the Apollo capsules through recounting some of the behind-the-scenes activities that took place during the 1960s as part of the Apollo program, which together helped make possible the success of Apollo 11. The story includes the program management, some of the engineers who designed the Apollo capsules and some of the people who helped to manufacture the Apollo capsules and some of the rocket boosters that culminated in the Apollo 11 astronauts landing on the moon in

1969. It also includes the contribution made by Hispanics to the design and manufacture of the Apollo capsules.

It is not unusual that most people do not realize that we Hispanics participated and contributed to the July 20, 1969, landing on the moon. They are also not aware that Hispanics continued to contribute beyond the initial moon landing, applying their engineering abilities to work on the subsequent lunar missions as well as the design of the Space Shuttle. The reason for this discrepancy is that there was a lack of advertisement and lack of media exposure about our contribution. I can attest that there were numerous talented Hispanics in America's space program because I was part of it and worked with several of them and knew many more through other projects that I participated in through North American Rockwell International. Most of us are familiar with the Apollo program and the Space Shuttle from televised programs and first-person accounts written by the astronauts, but we have never really heard the story about the people who made it happen—the engineers, the technicians, the managers, the executives, and the persons who built those fabulous machines.

Documentation exists of the moon landing and of events leading to it in many forms. There are magazines, books, and newspaper articles dealing with Project Mercury, Gemini, Apollo, and the Space Shuttle, and there are some excellent television programs presenting our space endeavors. While I applaud these programs, there is no mention of a Hispanic name on any TV space program. This book will include many Hispanic names and hopefully help us claim our chapter in history.

Neil Armstrong and Buzz Aldrin's landing on the moon on July 20, 1969, has been heralded as the most significant event of the 20th century, and we Hispanics who contributed to this great event should share in being part of it. The television media, of course, did a fabulous job of covering the dramatic event and was totally responsible for public awareness. It, however, neglected to make the public aware of contributions made by Hispanics. No one ever mentioned that the designer of the Apollo digital computer—the computer that guided the three astronauts from Earth to a moon orbit and then safely back to Earth—was Dr. Ramon Alonso from MIT, a brother Hispanic. Dr. Alonso accomplished this stroke of genius twenty years before the introduction of the PC. Just think for a moment: the computer that was the brains of the Apollo spaceship that took astronauts to the moon and back home safely was designed by a Hispanic man. It is the same computer that was used to guide the Lunar Module in its descent from lunar orbit to land the two astronauts safely on the moon and then launch the astronauts off the moon and guide the Lunar Module to rendezvous with the orbiting Command Module.

So, how critical was Dr. Alonso's digital computer? If you saw the movie *Apollo 13*, there was a scene that involved this computer. However, it was never referred to as Ramon Alonso's computer because it was called the CMC (Command Module Computer). In some earlier scenes, as Apollo 13 was headed for Earth reentry after the Service Module explosion, Tom Hanks (who played Commander Jim Lovell) was urging the ground controllers at Mission Control in Houston to supply the procedure and plan for Earth reentry. The reason they could not give Hanks the procedure was because if they told him that he had to turn on the CMC for reentry, it would suck so much battery power during the reentry flight phase that it would deplete the Command Module reentry batteries before they completed the maneuver, which would have been catastrophic for the capsule and the crew. The important movie scene is the one in which, down on the ground in Mission Control in Houston, Gary Sinise (who plays astronaut Ken

Mattingly) is lying on his back in the Command Module simulator, aided by Loren Dean (who plays John Arron), and they are trying to determine which noncritical systems can be turned off during reentry so that the CMC will have enough power to bring the Apollo 13 capsule all the way home to touchdown in the ocean. That CMC was Ramon Alonso's computer, and without it the Apollo 13 crew would have never made it home. For a normal Earth reentry, the fuel cells in the Service Module would power up the spacecraft for the first part of the reentry, but the SM was no longer available because of the explosion.

Earlier, I stated that the landing of the two Apollo 11 astronauts on the moon on July 20, 1969, was the most significant event of the 20th century. Nor am I the only person to take this view; the belief that it was the most significant event of the 20th century was expressed by several important and famous people, including the historian Arthur Schlesinger, Jr., who, when asked in 1999 what he thought the most significant event of the 20th century was, answered the Apollo 11 mission, the first landing on the moon.[1]

Because I made some statements previously about Ramon Alonso being the designer of the Command Module and the lunar lander computer, some words about him are appropriate here: When I was working on the Apollo program in 1964 to 1969, I was aware that there was a Command Module Computer for guidance, navigation, and control, but I did not know the details about the company that designed it or the computer designers. I came across the names J.H. Lansing, Jr., Albert Hopkins, Richard Battin, Ramon Alonso, and Hugh Blair-Smith[2] as the early designers of the architecture of the Apollo computer. The name of the Hispanic designer, Ramon Alonso, popped out at me. I found out that Ramon Alonso had received his PhD from Harvard, that he had prior experience in designing small aerospace computers (70 pounds), and that he used rope memory borrowed from the Australians with 64 KB of ROM (read only memory) and 8 KB of RAM (read and write memory) programmed in Assembly Language, and the 70 pounds were packed in 24 by 12.5 by 6.5 inches.[3]

This book was written based on behind-the-scenes work in which I participated during the Apollo program from 1964 to 1969 and from personal association with many of the events that led to the moon landing, as well as contact with the engineers, supervisors, managers, executives, and Hispanics who helped us get to the moon. Those of us who contributed to the events leading to the Apollo 11 mission and to the actual landing on the moon itself feel a deep pride and gratification for being part of that endeavor. Although the real feeling can never be described, I hope that as readers go through these pages, they get a sense of what it was like to work on a project such as Apollo.

Some of my Hispanic colleagues who worked on the Apollo program at Rockwell International include Manny Adame, Angel Aguilar, Jake Alarid, Jose Alvarez, Manny Alvarez, Manny E. Alvarez, Ester Beaty, Oscar Buttner, Sandy Cardona, Leonard Colacion, George Cortes, Jorge Diaz, Roman Dominguez, Ernie Fuerte, Tony Gaitan, Frank Gamboa, Bert Garcia, Gilbert Garcia, Gil Garcia, Paul Garcia, Ted Garcia, Joe Gomez, Ralph Gomez, Marty Gutierrez, Val Guzman, Jose Jaramillo, Tom Lopez, Bob Loya, Larry Luera, Hector Maldonado, Hank Martinez, Al Mejia, Marty Mestas, Herb Montano, Herb Montoya, Cruz Mora, Danny Moreno, Elisa Munoz, Pete Ontiveros, Phil Padilla, John Perez, Ben Reina, Reynaldo Reyes, Fred Rodriguez, Bill Ruiz, Joe Salazar, Ray Sena, Frank Serna, Bob Solis, Manuel Talamantes, Tony Vidana and Frank Vigil. There is no doubt that there are many others not mentioned here, but some names of employees whom I never met do appear in this book. I know there were many more

Hispanics at Rockwell because, while I was working on the Apollo program, many Hispanics whom I did not know were featured in the company newspaper.

During the turbulent 1960s, there were at least three Hispanic organizations at North American Rockwell that worked for the benefit of Hispanic employees and Hispanic students; one was Youth Incentive Through Motivation (YITM), another was the Chicano Forum headed by Larry Luera and the third one was the Hispanic Aerospace Workers Consortium. Some members of these organizations whom I was not privileged to meet were Tony Barron, John Camarena, Charlie Chavez, Gilbert Contreras, Raul Espinoza, Dave Gutierrez, Ed Gutierrez, Bill Ponce, Tony Saldate, Ike Garcia, Ernestine Hernandez, Phyllis Jaime, Lizzie Johnson, Margaret Saenz, Laura Robles Lopez, Lucy Vega, Pete Moraga, Luciano Aguirre, Pete Barrera, Joyce Suzuki, Candelaria Terrazas, Rick Villasenor, Lourdes Villicana, Jerry Gerard, Bob Jimenez, Gene Silvas, Connie Suggs, Rosa Flores, Louie Mendoza. Delia Marquez, Walter Rivera, Cruz Padilla, Nellie Leyba, Phyllis Perez, Richard Manriquez, Abigail Bahktiari, Lupe Caro, Narcisso DeLeon, Elizabeth Garcia, Paul Martinez, George Torres, Rebecca Trujillo, Roy and Rosemarie Mungaray, and Mary Lou Garcia.[4]

Some of those I named in the first of the above two paragraphs were my colleagues at work; others were only members of YITM, the organization of which I was temporary chairman when the organization was founded (later I became its first president for its first year of operation). Of those persons I mentioned in the second of the two paragraphs, I personally knew Larry Luera because I attended his Chicano Forum lunch meetings many times. Later, after I was no longer working for North American Rockwell, I met Rick Villasenor at a meeting at his house on May 3, 2018; Rick had been president of the Hispanic Aerospace Workers Consortium.

In the old Aztec Empire, before the Spanish conquest, one of the highest ranks that an Aztec warrior could achieve was that of the Eagle Knight. So, several of the chapters in this book refer to "Eagle Knights" in their titles because if Hispanics mentioned here had been members of the Aztec Empire, then, by virtue of their achievements, I believe they would have been knighted as Eagles.[5] My reference in this book to Hispanics who helped put humans on the moon as Eagle Knights is not far from the mark because Mexico's highest award bestowed on foreigners is the Order of the Aztec Eagle (in Spanish, *La Orden Mexicana del Aguila Azteca*). Among those to receive this award in recent years is Jared Kushner, as a result of his contribution to the NAFTA negotiations.[6]

Introduction

To this day, there are still many skeptics who do not believe that the United States landed humans on the moon. These nonbelievers are not mathematicians, engineers or scientists, so why should we acknowledge them and believe them instead of those who designed the vehicles and the systems that took astronauts to the moon? Of course Neil Armstrong and Buzz Aldrin really set foot on the moon, as did ten other astronauts. It was possible to journey to the moon in the 1960s and 1970s because there were well-known scientific laws and mathematics and engineering technologies that made this endeavor possible, but there were also people with the necessary vision, including Soviet engineer Yuri Kondratyuk, Dr. Robert Goddard, Dr. Wernher von Braun and John F. Kennedy—without these four men, the landing on the moon might not have happened.[1]

To begin with, there were the mathematics that made possible the fundamental sciences and engineering technologies that are necessary for space travel. So, what is needed for humans to travel from one celestial body to another, such as from Earth to our moon? First, you need a vehicle to get you there and the engineering technology to survive the trip. You also need to know how far the moon is from Earth and where the moon is in its orbit around the Earth at any moment. In addition, you must have some knowledge of celestial navigation, some space guidance knowledge technology to get you there and back. Equally necessary is some device on the vehicle that can orchestrate the whole trip, keep track of where the vehicle is during its journey and receive commands from the crew. The vehicle must not be too big and too heavy, and all the equipment inside the vehicle must be small and able to withstand the harsh environments of the launch and temperatures of deep space. Since the trip we are talking about is to the moon, the crew will also need a space suit to keep them alive in space, as well as a means of wireless communication. And let us not forget a planned path for the total journey to get the space vehicle to the moon and return it safely to Earth. So, let us see whether all that existed in the 1960s.

The invention of the vehicle to get humans to the moon started in 1907 with Dr. Robert H. Goddard, the father of rocketry, whose work continued until 1945. His theory and inventions made the firing of rockets and missiles possible. Further development of rocketry was spearheaded by the German scientist Dr. Wernher von Braun, who successfully developed the V-2 rocket during World War II that was used to bomb England (V-2 rockets were launched vertically from a platform out of the Peenemunde Army Research Center in Germany). After the war, von Braun and some of his best engineers and scientists were brought to America to work on the space programs Mercury and Gemini. Dr. von Braun told American scientists that his work on the V-2 rocket was highly influenced by the pamphlet "A Method of Reaching Extreme Altitudes" by Dr. Goddard.[2] By the early 1960s, von Braun and his German engineers and scientists stationed in Huntsville, Alabama, had developed on paper a heavy-lift booster capable of taking astronauts to the moon. It must also be pointed out here that all the mathematics and science necessary for Dr. Goddard and Dr. von Braun to do their work were already available. There was the theory of gravity discovered by Isaac Newton (who also invented calculus); aerodynamic theory used by the Wright brothers for the invention of their airplane and further developed for military aircraft flight; the science of thermodynamics, which allowed engineers to understand the relationship between friction and heat; chemistry (the successor of alchemy), which allowed the development of liquid and solid propellant fuels for the vehicle engines; and the theory of electronics, which was sufficiently developed to allow Dr. Goddard and Dr. von Braun to design and build any electronic system they required. The development of the transistor and the digital computer came much later, but they were available in the 1960s. It was solid-state physics that brought about the invention of the transistor, a solid-state semiconductor device that paved the way for the development of more solid-state devices that could withstand the high vibrations of the rocket launch environment and replace vacuum tubes that could not withstand such conditions. Solid-state devices also made possible the invention of the robust small digital computer that allowed engineers to minimize the size and weight of the systems that went into the Apollo vehicles that took men to the moon and returned them safely back to Earth.

The two large space programs that the United States spearheaded for several years, Mercury and Gemini (launched into space from 1959 to 1966), led to the development of the space suit, the heat shield for the spaceship and its crew to survive the heat of reentry, and the technology for humans to survive in outer space. The two programs also provided astronauts with an opportunity to learn how to maneuver spaceships in outer space and to experience the effects of weightlessness. As for the knowledge of the positions of the Earth and moon at any given time and their relative positions to one another, such knowledge has a long history that goes back to the Greeks and later astronomers like Ptolemy, Tyco Brahe, Johannes Kepler, Isaac Newton, Nicolaus Copernicus, Galileo and many others who documented all their fundamental discoveries (which I learned about in the astronomy course I took at UCLA in 1949 and served me well in my chosen career).

So, how did we get to the moon and back to Earth by the end of the 1960s? It started with navigation. By the 1960s, ships at sea and aircraft had been navigating around the planet for centuries using the stars and the earth sphere in conjunction with the sextant, an instrument used by navigators to ascertain their position on Earth by sighting heavenly body positions and the Earth's horizon. To perform celestial navigation, the earth sphere was extended to a celestial sphere to develop celestial navigation using star sightings. Better yet, submarines and missiles had been navigating with inertial navigation systems and the Kalman filter for years, and that system was certainly available for a trip to the moon in the 1960s. As for guidance techniques, NASA sent and soft-landed unmanned satellites on the moon from 1966 to 1968 during the Surveyor program with the aid of Jet Propulsion Laboratories, Hughes Aircraft Company, Draper Labs at MIT, and many other companies too numerous to mention. The Surveyor moon landers gave us guidance and navigation techniques to get to the moon and the ability to soft land on the moon, which was all needed for a manned moon landing. To return humans from the moon, all we had to do was reverse the Surveyor guidance and navigation type of techniques. In the 1960s, Draper Labs at MIT was developing guidance equations and techniques for a trip to the moon. The orchestration of the Surveyor spaceships was accomplished by a digital computer that had its early development in the EDVAC computer at the University of Pennsylvania and the ENIAC computer. Such great mathematicians as George Boole gave us Boolean algebra for the logic of digital computers, and the great mathematician John von Neumann provided the computer stored program and its instruction set. For wireless communication, we used the Marconi invention, which had been highly developed by the 1960s.

Having all these inventions and technologies at their disposal, engineers in the Apollo program had the task of putting them all together and building a spaceship that could safely carry three crewmen to the moon and back to Earth. The task was not easy, and our Apollo team worked hard and made many sacrifices to put three astronauts on the moon by the end of the 1960s. The United States of America went to the moon not because it was easy but because it was a challenge, and someone had to go, so why not us? How difficult was it? Well, we were essentially shooting at the moon, which is a moving target, from the Earth, which is a moving celestial body. Remember that we were shooting a rocket to the moon, a rocket that would take several days to get there, so the trick was to shoot the rocket to a point in space where you had calculated that the moon would arrive at the same time when the rocket arrived.

We did not just one day decide that we should go to the moon. The desire and plan of going to the moon started many years ago, in the 1920s.[3] Working on a program such as Apollo was challenging and interesting but not glamorous. As you read the various chapters of this book, you will find much detail, because that is what it took—much attention to detail, leaving little to chance. Every action that was undertaken during the Apollo program had a probability of succeeding or failing, and engineers strove to maximize the probability of succeeding and minimize the probability of failing because the Apollo program had to succeed to win the space race; we had to have the crew survive the mission because an accident in space and failure was not an option. The whole world (and especially the Russians) was watching us, and America's prestige was on the line.

All names in this book are real names of real persons in the Apollo program. Also, as you read this book, you can see how the career of an engineer matured to be able to solve complex problems and to lead teams to move a project forward. The book will also bring awareness to how supervisors, managers and high-level executives operated to make a large, complex program like Apollo successful. This book will illustrate some of the things that engineers do, ideally creating interest and inspiring young students (especially Hispanics) to pursue a scientific or engineering career.

1

The Apollo Unmanned Capsules and Some Eagle Knights

Close to the end of March 1964, the research and development activity on the Hound Dog cruise missile at North American Aviation Space Division in Downey, California, which I had been working on, had been completed, and the manufacturing of missiles to deliver to the Air Force was going into full swing. At that time Ed Kelley, supervisor of the research and development Hound Dog Cruise Missile Flight Control System Design Unit, stopped by my desk and told me that he wanted to talk to me in his office. When we both sat down, Ed said, "The Hound Dog cruise missile manufacturing is being transferred to Tulsa, Oklahoma, and the powers that be had fingered you to go with it to support the manufacturing line. I told them that you were a research engineer, not a manufacturing engineer, and that it was a bad idea, that you were more valuable to the company if you stayed here to work on any one of the ongoing research programs. They heard my plea, and so you have my permission to find a job in any of the company's ongoing research programs." The Hound Dog cruise missile was the first cruise missile to be fielded into active duty by the Air Force. Because I was a senior research engineer, it was a relief to me that Ed had interceded on my behalf and had given me permission to find employment in any of the ongoing research programs. At the time, my wife was seven months pregnant with our third child; she was prone to miscarriages and heavily medicated, and thus in no condition to travel, so I have no idea how I would have handled it if we had been obliged to move to Oklahoma.

After Ed had given me the news that I could go look for a new job within the company, I had no doubt that I could get one, but would it be a job with a challenge that I would enjoy and that had a promising future? I knew several engineers who were working on Apollo; two of them were Ken Watson and Paul Garcia, who was the first Eagle Knight I had met when I first started to work for North American Aviation in 1960. I also knew Claire Harshbarger, who had been one of my supervisors and was now chief of the Para-glider Flight Control System, and Will Owens, who was a project engineer under Claire.

There was also Bob Antletz, a supervisor who had hired me to work on the Hound Dog cruise missile in 1960. Bob had worked on the Saturn II (Saturn Two) booster rocket but was now working on the Apollo program. I could get names of supervisors and managers on the Saturn program from Bob because he knew Bill Parker very well. Bill Parker was a company vice president and program manager of the Saturn program. Bob had told us one day while he was still on the Saturn program that Bill Parker had pulled him off the Hound Dog cruise missile program in 1961 to have him work on the Saturn program. Bob also said that all the time he was there he hated it, because as far as he was concerned the Saturn II booster was nothing more than a big stove pipe. He wanted to leave that program badly and told Bill repeatedly that he wanted out. He went to see Bill in his office one day to seriously confront him and asked him when he could get off the program. According to Bob, "When I asked him that, Parker got really mad. He looked at me with fire in his eyes." Then Bill Parker said to Bob, "You will get off of this program when I say so." Bob said he told Bill, "Who do you think you are, God?" Some weeks after that encounter, Bob was off the Saturn program and on the Apollo program. Bob was now systems engineer for the Apollo Stabilization and Control System for the manned Apollo capsules. However, I had no plans to call Bob Antletz for names of supervisors and managers who worked on the Saturn program because the prospect of working on the Saturn program was not something that particularly interested me.

Some background information is in order here because you are probably wondering, who are those persons mentioned in the last two paragraphs? So, let me recap some earlier events to put things in perspective: After serving four years in the Air Force Training Command during the Korean War as instructor of electricity and radar electronic circuits, I went to work for Hughes Aircraft Company in Culver City, California, on Interceptor Aircraft Radar Systems. While attending the last semester of my senior year at USC, due to company changes, I was forced to find employment elsewhere. In February 1960, Bob Antletz interviewed me for a job as a research engineer at North American Aviation. During the interview, I told Bob that I would accept an offer of employment only if I could finish the semester day classes at USC for my BSEE degree and work eight hours after my last class each day. Bob told me that he would review my resume and give my proposal some consideration. Several days after this interview, North American Aviation Human Resources called me at home and told me that Bob was making me an offer of employment that was based on my work proposal to him so I could finish my studies for my degree. I accepted the offer, and on my first day at work Bob assigned me to work for Ed Kelley, who was working on the cruise missile flight control system. Ed became my mentor and eventually my supervisor. Claire Harshbarger was Bob Antletz's design unit lead engineer, and Ken Watson was a senior engineer in Bob's design unit. Paul Garcia was a test engineer at the Downey facility where Research and Development Hound Dog cruise missiles were processed with added instrumentation so that the company test team in Florida could launch the missiles to prove to the Air Force that the cruise missile met its required capabilities. One of the things Ed Kelley told me was that the engineers named in this paragraph (except for me) were all World War II veterans and had prior experience working on rockets because they were part of the North American Aviation team that worked on the Navaho program, which had started research activities in 1946. The Navaho consisted of a cruise missile named the X-10 that rode piggyback on the Navaho rocket (based on the German V-2 rocket). The contract was cancelled in the late 1950s in favor of ICBMs as a first line of nuclear defense instead of cruise missiles.[1]

I made some phone calls to Ken Watson, Paul Garcia, and others, from which I learned that Dr. John Kalayjian was supervisor of the Apollo Stabilization and Control System Design Unit, and Bill Fouts, who had been lead dynamist on the Hound Dog cruise missile program, had just recently been appointed chief of Apollo automated systems for unmanned Apollo capsules and was forming new groups. On phone calls to John and Bill, I made appointments to interview with them. The first interview was with John Kalayjian. During the interview in John's office, we talked for a while, but the interview turned out to be a session for John to tell me all the things that he had done, and we never really talked about my capabilities and how I could help John's design unit. The second interview was with Bill Fouts. Bill knew me personally and my reputation as an engineer on the Hound Dog cruise missile program, and his assignment as chief of automated systems on unmanned Apollo capsules seemed like an especially interesting one with plenty of challenges and opportunities. Because his was a nascent Apollo contract, he was hiring and said that I had a job with him if I wanted it.

Since John Kalayjian never acknowledged that he needed me during our interview, and because John had all his lead jobs filled, I was not too eager to join his group even though working on the Apollo Stabilization and Control System was what I would have preferred to do. I was afraid that working for John could end up being some meaningless job. As badly as I wanted to work on the design of the Stabilization and Control System for the manned Apollo capsules, I relented and sometime in early April called Bill Fouts and told him I would take the job on the unmanned capsules. I knew that on the unmanned Apollo project there was at least a good chance of being a lead engineer. A transfer request was typed for me by the Hound Dog management on April 15, 1964, and made my transfer to Apollo effective on May 17, 1964. The transfer was signed by Sandy Falbaum, Russ Belles, and Al Dale on the Hound Dog side and Charlie Feltz, Gary Osbon, Dave Levine, and J. McCarthy, Jr., on the Apollo side. I reported to Building 6 in Downey prior to my effective transfer date. There was no problem with having the transfer approved because I personally knew Sandy Falbaum, who had been assistant program manager, and Al Dale, who had been project engineer on the Hound Dog side, as well as Gary Osbon, who had been chief engineer on the Hound Dog cruise missile and was now chief engineer on the Apollo program.

On arrival at Building 6 of the North American Aviation Space Division facility, I was taken to an area that had quite a few desks but very few people and was told to take any desk—the groups had not yet been organized and probably would not be for quite a while. Finally, I was on the Apollo program, which was at the leading edge of space technology and the center of the space race against the Russians, which was no small thing. It was a new department just being formed, but I was not sure exactly what it was that I was getting into. After selecting a desk, I sat down, looked around the immediate area and noticed that there were no recognizable faces. With no instructions about what to do and lacking a list of duties (other than those that had been typed on my transfer request), I decided not to sit there for one, two, or three weeks, or even one day, and do nothing. I decided that it was best for me to get some reading material so I could start boning up on the Apollo program.

For those of you readers who are too young to know the details of the Apollo program, let me tell you what was going on in the program at this time in 1964 and before. To understand the function of Bill Fouts' new Automated Systems Department, you must understand the Apollo program and its early development. In the 1950s, Wernher von Braun and his German rocket engineers and scientists who were brought over to the United States from Peenemunde, Germany, after the end of World War II were in Huntsville, Alabama, designing and building the rockets that would launch the Mercury and Gemini capsules, which grew out of their earlier work on the V-2 rocket in Peenemunde. Also, in their free time in Huntsville they were designing a booster rocket big enough to launch a capsule to the moon. They were doing this work because von Braun was a visionary and had dreamed since he was sixteen years old of someday sending astronauts to the moon. However, the group had to work on the moon-bound booster design on their own time because funding for such a project had not yet been authorized.

After President Kennedy's announcement in 1961 that the United States was going to the moon before the end of the decade, von Braun's team was eventually given the funds to legally work on the large booster project. The booster was given the name Saturn V (Saturn Five). During this same period, there were three possible methods being debated regarding how to get astronauts to the moon and safely back to Earth. This debate involved NASA managers in Washington, NASA managers at the Manned Spacecraft Center in Houston, and von Braun's team in Alabama.

The first method was called the Direct Mode. The Direct Mode required a rocket that would lift off with its payload and head in a trajectory directly to the moon and then land on the moon. The return trip would likewise be a direct trajectory back to Earth. This mission was favored by many because it was uncomplicated, requiring no rendezvous and no mating between two space vehicles. It did, however, require twice the lift-off power of the Saturn V rocket booster because it would carry the fuel and payload for the trip to the moon and fuel and payload for the return trip to Earth from the moon. The booster assigned to this Direct Mode mission was the Nova booster, a project that was going to be under American development but had not yet started, as opposed to the Saturn V, then being developed by von Braun and his German team of engineers. What made a large booster like Nova necessary for this type of mission was that going directly from the Earth to the moon would require a considerable amount of fuel, more than Saturn would be able to provide. The opponents to the Direct Mode believed that the Nova booster development could not be completed by the end of the decade, thereby jeopardizing President Kennedy's timetable of getting Americans to the moon in less than ten years.

The second moon mission proposal was known as the Earth Orbit Rendezvous. This mission required two Saturn V boosters to be launched. One Saturn booster would lift into Earth orbit carrying the vehicle to be used for moon travel, moon landing, and a return trip to Earth. The moon travel/landing vehicle would contain no fuel. The second Saturn would lift off with a tank full of fuel into the same Earth orbit as the moon travel vehicle. Once both were in Earth orbit, the two vehicles would then rendezvous, and the moon vehicle would be filled with fuel for the moon trip and return mission from the tank vehicle. The lunar travel vehicle would then head for the moon and return to Earth in the same way as the Direct Mode. On reflection, it becomes obvious that this mission was a modification of the Direct Mode in which the heavy payload would be divided between two Saturn V boosters being launched into Earth orbit, where they would then combine to form one moon travel vehicle for a direct trajectory to the moon and back to Earth. The proponents of this type of mission favored it because it did not require the development of the Nova booster. It would, however, require a considerable amount of fuel for the direct trajectory both ways.

The third option was the Lunar Orbit Rendezvous method. This mission required one big launch booster such as the Saturn V to be launched into Earth orbit carrying the lunar travel vehicle and a lunar landing vehicle with all the required fuel for the mission. The lunar travel vehicle would consist of a capsule that would serve as the habitat for the astronauts while on the journey to the moon, as well as a lunar landing vehicle that would allow two astronauts to land on the moon. Once in Earth orbit after the launch, the moon travel vehicle and the moon landing vehicle attached to it would then transition into a trajectory toward the moon. On reaching the moon, the vehicle would shift into a moon orbit. From lunar orbit, the moon landing vehicle would then descend to the moon surface, with the capsule that it was attached to remaining in orbit around the moon with one astronaut left behind to man the capsule. The lunar landing vehicle carrying two astronauts would consist of two vehicles with separate engines (one engine for landing on the moon and the second for blasting off the moon to return to the capsule left in moon orbit). When the mission on the moon was completed, the lunar landing vehicle would blast off the moon with the second stage, using the stage that it used to land on the moon surface as a platform, which would be left behind on the moon. The moon blastoff vehicle would then transition into lunar orbit followed by a rendezvous with the moon orbiting capsule. The attached combo would then transfer to a trajectory toward Earth. The moon blastoff vehicle would be ejected into space once the lunar traveling astronauts had transferred back into the capsule. The capsule housing the three astronauts would reenter Earth's atmosphere, deploy parachutes and land in the ocean. This was the most complex of the three mission plans but the one that used the least amount of fuel. The rendezvous maneuvers were the most complicated aspects of this mission, but they were rehearsed and perfected during the Gemini program, so there was minimum risk there.

The three biggest objections to the Lunar Orbit Rendezvous method were the mission's complexity, the use of the German-designed Saturn V booster, and the fact that this plan was not developed in America. The Lunar Orbit Rendezvous mode was first suggested by Ukrainian rocket pioneer Yuri Kondratyuk in 1919 and later proposed by von Braun in his book, *The Mars Project*, in 1962. After much debate and politicking back and forth, and putting all jealousies aside, the Lunar Orbit Rendezvous mission was chosen as the method for reaching the moon by the end of the 1960s because it had the best chance of success. This then became the Apollo program. The history that I have written here is paraphrased and simplified so the reader can understand it because it took a long time to make a final decision because it was a very complex problem.[2]

The original plan for Apollo was to design, build and test all the moon mission vehicle sections on the ground, with minimal testing accomplished in space, and then launch three astronauts to Earth orbit followed by a mission to the moon and back to Earth. Eventually people at NASA and at North American Aviation started to wonder whether it would be wise, for safety reasons, to send some unmanned Apollo spaceships out into space to test them before proceeding with manned missions. Sending astronauts into space in newly invented vehicles, with minimal previous space testing, to a place where we had never been before was, to say the least, a very risky endeavor. The first change to the original plan came in 1962, which involved adding one Apollo unmanned capsule to be launched into a ballistic trajectory and fly a suborbital mission to test the capsule and some of the Apollo systems and space travel techniques. The capsule for this mission was dubbed Airframe 9 (always written as AFRM 009), and the automated system that was to fly this space capsule was called initially "Mechanical Boy." This was the task given to Bill Fouts, and this is what I had come to work on. The mission of AFRM 009 was designated Mission SA201 by NASA.

There were three more Apollo capsules that were going to fly unmanned, but the capsules and their mission had not been determined at this time. However, later, while work on AFRM 009 was in progress, NASA and North American Aviation selected AFRMs 011, 017 and 020 to be the other three Apollo capsules to fly unmanned missions as a prelude to Apollo manned orbital flight and the moon mission. There will be more to say about these three airframes because there was no way that Bill Fouts' organization could escape being involved with them.

The test objectives of Mission SA201 for AFRM 009 were defined in a letter from Chief Engineer Gary Osbon that had an original release date of February 16, 1962, then was revised on September 26, 1963, and revised again on May 25, 1964. This letter designated AFRMs 009, 011, 012, 014, 015, 017, and 020 as a Block I concept. The Block I designation meant that this group of vehicles would be identical and of the original design and to be qualified for Earth orbit missions only because their design was not suitable for the long journey to the moon and back. Any design changes stemming from their test results and any other design changes defined as necessary for a successful moon mission would be grouped together to form a new airframe design, which would then be designated Block II Apollo capsules. The major change to Block II Apollo capsules started in June 1964 when the Block I Stabilization and Control System was redesigned to the Block II as a backup to Command Module Computer being designed by Ramon Alonso at MIT.[3]

On a moon return trip, a manned spacecraft had to stay within a predetermined tubular corridor that got narrower as it approached Earth; the real dangers here were that the capsule could stray from the corridor and come into Earth's atmosphere too steep and burn up or else come in too shallow and skip out into space, in both cases losing the spacecraft and its three-man crew. It was the job of the onboard guidance and navigation system, aided by the ground control systems and the guidance and navigation techniques, to keep the Apollo capsule inside the corridor. The real question was whether the Apollo ground control systems and the guidance and navigation techniques could hold the spacecraft within the corridor from the moon all the way to Earth. These systems and techniques had never been tested in a real situation in space, and this had to be done because proper operation of this system in space was mandatory for the success of the Apollo moon mission. One of the main objectives of the last three unmanned Apollo capsules was to certify some of these systems for a moon mission. In addition, at this time, neither the Russians nor the Americans had attempted a soft landing on the moon by a space vehicle. A soft-landing capability on the moon was a requirement for a successful Apollo moon landing by 1969. The American plan was to acquire this capability with the Surveyor unmanned moon landing vehicle by the mid–1960s. There were various documents that described the Apollo entry corridor when I worked on the Apollo program from 1964 to 1969, but in September 1969 I was assigned to work on the Space Shuttle and had little opportunity to copy some of those documents, so I wrote this corridor description from memory.[4]

When returning from the moon, the Apollo capsule would be reentering Earth atmosphere much faster than when reentering from Earth orbit, and so the reentry would be much, much hotter; the capsule thus required a better heat shield than what was used on the Mercury and Gemini capsules. The Apollo heat shield now had to be tested and qualified for a moon reentry velocity if the Apollo crew were to survive a moon mission. It was also possible that some of the systems meant to keep the astronauts alive in the freezing coldness of space would not work right, and the Apollo crew would freeze to death and be lost in space forever on their way to the moon or during their return to Earth. For these reasons AFRMs 011, 017 and 020 were planned to fly unmanned. As in AFRM 009, in place of the astronauts there was to be an electronic automated system that would automatically perform all the functions that an astronaut had to perform to fly the space capsules. The electronic automated system for these three unmanned spaceships was originally called "Mechanical Man." The missions for the three additional unmanned capsules were more complicated than the AFRM 009 mission, and so the Mechanical Man was more complex in design than the AFRM 009 Mechanical Boy. It had complex digital logic to be able to make decisions based on the information it received and then send out commands to the other systems that it was designed to command. In its electronic complexity, it tended to be more like a hardwired minicomputer rather than a simple system like the Mechanical Boy.

On my first day in Building 6 it was imperative for me to visit the Block I Stabilization and Control System Group and talk to Paul Garcia and Ken Watson to get as much information as I could from them regarding the Apollo program. When I visited them, Paul gave me a schematic drawing of the Apollo Block I Stabilization and Control System that he said was the system to be installed in AFRM 009. Putting the schematic together for the Block I Stabilization and Control System was Paul's responsibility, and he had access to extra copies. Paul briefly described the operation of the system to me using his schematic

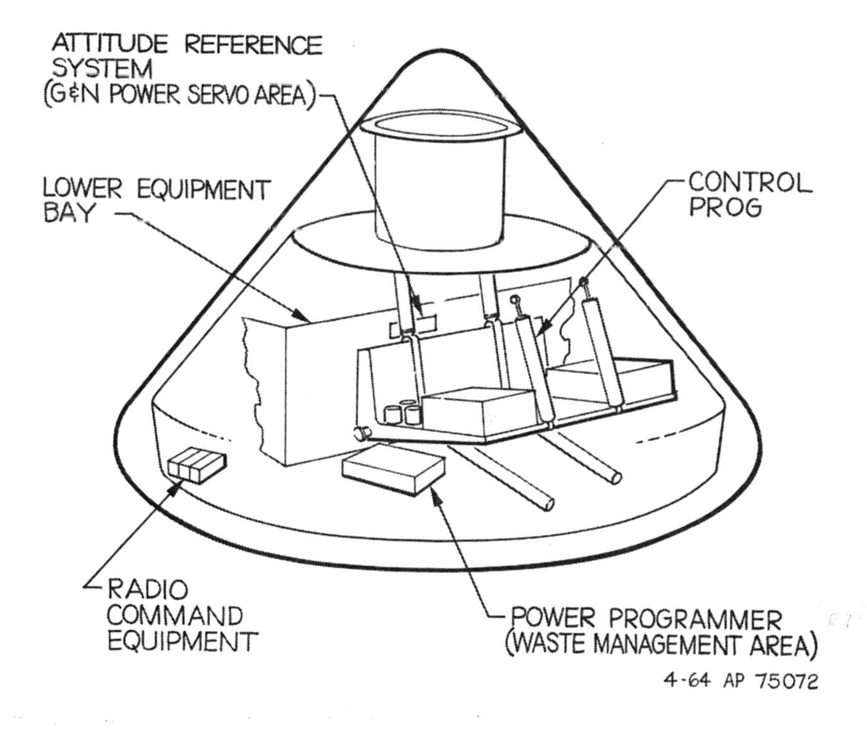

This schematic shows the Control Programmer and Attitude Reference System installation on the AFRM 009 platform (Courtesy of the Boeing Company).

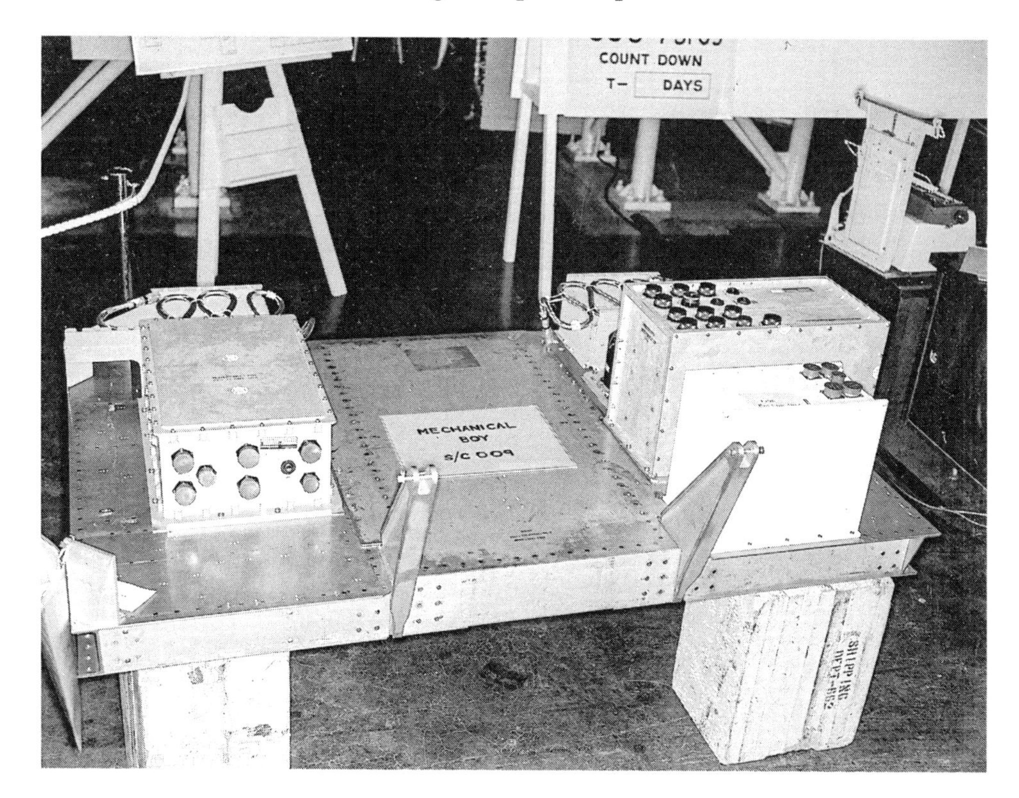

The two large boxes are part of the Control Programmer on the AFRM 009 CP platform. Part of the black Lockheed timer is seen behind the big box on the right side (Courtesy of the Boeing Company).

drawing. Paul also gave me several documents that described the system in more detail. Both Ken and Paul gave me other material to read along with names of important players in the AFRM 009 project who could supply more reading material. While I was there, Paul introduced me to Lloyd Campbell, George Cortes, Mert Stiles, Pete Ontiveros, Larry Pivar, Vester Purkey and Danny Moreno. The only one of the seven whom I knew from previous encounters was Vester Purkey because I had met Vester at Eglin Air Force Base in Florida during the time I was there supporting the Hound Dog cruise missile flight test research program. Now I had met three more Eagle Knights: George, Pete, and Danny. Paul, Danny, and I talked for a few minutes and then said our goodbyes. Later I collected every type of material I could get my hands on from the different persons whose names Paul Garcia and Ken Watson had given me. Later, I read everything as rapidly as I could and knew the whole story of Apollo and AFRM 009.

After visiting Paul and Ken and reading all the material I was given, I was extremely excited, as well as nervous, because my dream of being part of the team that was responsible for getting three men on the moon was finally coming true. My gut told me exactly what had to be done to move the project forward, but there was a need to gain some more knowledge of Apollo and Mechanical Boy before confronting the Mechanical Boy designers and other engineers on AFRM 009. Even if the design of the AFRM 009 mission and the Mechanical Boy had already been started, there was always a period early in any research and development project when mission changes and redesign of systems took place. It was almost a sure bet that this was going to happen on AFRM 009.

I spent some time reading Apollo material at work for several days while the empty desks kept filling up slowly. I wanted to be in a position like the one I had enjoyed on the Hound Dog cruise missile but wanted more authority to make decisions that contributed to getting our astronauts to the moon by the end of the decade. This was the ground floor of a new project and a new department, so the opportunity for all my wishes to come true was there.

It is important to know how the Apollo management and departments were organized and what hardware of the Apollo program each department was responsible for designing, developing, building, testing, and delivering to NASA. Remember that in 1964 at North American Aviation none of the moon-rated spaceship components had been completely built, even though some Apollo space capsules were being constructed on the manufacturing floor, while others were still in the process of being developed. The following description details the Apollo rocket concept and its various components that were required for a moon mission: By 1964, companies had already been awarded contracts to design, develop, build, and test the pieces of the Apollo rocket. Starting from the bottom up, the first section was the Saturn V booster that was on contract to Boeing—the most powerful booster ever to be built. The Saturn V booster was a long-term development project that was needed for the eventual moonshot but not for some of the earlier missions, such as AFRM 009. The second stage of the Apollo rocket was the Saturn II (Saturn Two), and the third stage was the Saturn IVB (Saturn Four B). On top of the Saturn IVB was the Lunar Excursion Module, which was covered by a shroud and was being designed and built by Grumman Aerospace in Long Island, New York. Attached to the Lunar Excursion Module was the Service Module that was being designed and built by the North American Aviation Missile Division in Downey, California, which was followed by the Apollo capsule, also a North American Aviation project. The Apollo capsule was a teardrop-shaped capsule with the blunt end facing downward and the apex end facing upward. Attached to the apex of the capsule at the very top of the rocket was a long, slender, solid rocket motor called the Launch Escape Tower, which was part of the Launch Escape System; the purpose of this solid-fuel rocket motor was to propel the capsule and astronauts to safety, away from the rocket, while on the launch pad or in the early part of the lift-off in case something went wrong and caused the rocket to explode. The Launch Escape Tower was to be jettisoned once the moon rocket was on its way and had cleared the launch pad safely. The Lunar Excursion Module, the Service Module and the Apollo capsule were the three pieces that were going to make the trip to the moon and were crucial for the moon trip, as was the CMC (Ramon Alonso's computer, discussed in the preface).

North American Aviation had the contracts from NASA for the Apollo capsule (or Command Module) and the Service Module, which were being designed and built in Downey, California. The other section of the Apollo rocket that North American Aviation was contracted to design and build was the Saturn II, which I mentioned earlier as Bob Antletz's hated stove pipe. Because I was not involved with the development of the Saturn II, I cannot comment on its design and development. However, I may refer to it frequently because of its involvement in the AFRM 009 mission and the moon mission.

At North American Aviation in Downey, Apollo hardware was divided into systems. Some of the systems were the Electrical Power System, Communications System, Environmental Control System, Vehicle Structures, Guidance, Navigation and Control System, Thermal Protection System, Reaction Control System, Service Propulsion

The Apollo Saturn V rocket; the author, in a white shirt, can be seen underneath the "U" in "USA." On the left is Saturn II (Bob Antletz's stove pipe). Picture taken at Kennedy Space Center in 1995 (author's collection).

System, Automated Systems, Entry Monitor System, and many others. Each system was assigned a systems engineer who was responsible for the technical end of the system. At the time that the AFRM 009 Automated System was being developed, the Apollo program had eighteen systems engineers, which meant that there were eighteen systems. Bob Antletz was systems engineer for the Stabilization and Control System, Nick Glavanich was systems engineer of the Reaction Control System and John Kalayjian was acting systems engineer of the Automated System. There was an engineering department for each system, which was headed by a manager. Under each manager were several supervisors, the number depending on how many were needed to complete the engineering task for his assigned system. Each manager also had a designated subsystem project manager and an assistant manager if required. In addition, managers had one or two staff people to help them in their administrative duties. Managers in the 1960s were all male because engineering was the domain of men, and, as I recall, there were no female managers at that time, but there were one or two female supervisors. A group of managers was then assigned to report to a director who in turn reported to the chief engineer. Above the chief engineer were the various division vice presidents, one of whom held the position of Apollo program manager and was the one to whom the chief engineer reported directly. All division vice presidents reported directly to the division president. The division president in 1964, while I was working on Apollo, was Harrison Storms, a graduate of Northwestern University who had a master's degree in mechanical engineering and then went to the California Institute of Technology and studied under Dr. Theodore von Kármán for a master's in aeronautical engineering. His nickname was Stormy, and because he was responsible for having NASA award the Apollo contract to

A closer view of the Saturn V Booster engine nozzles that illustrates their immensity (author's collection).

North American Aviation, he was known at North American Aviation as the father of Apollo. Stormy had a great management team that helped him secure the Apollo contract, and that team was nicknamed the Storm Troopers.[5]

The pressure from NASA would be directly on Bill Fouts because, in terms of the Apollo schedule, Bill's project was already running late: in addition to AFRM 009, four more Apollo unmanned airframes had been added to his program, but the final date for going to the moon had not changed. At this time in 1964, Joe Dyson was manager of the Guidance, Navigation and Control Department and reported to the director, Dave Levine, and Bill Fouts, as chief of a department, also reported to Dave Levine. Someone at North American Aviation in 1964 told me that Dave Levine was a management member of Stormy's Storm Troopers, and at that time I believed it because I had occasion to work closely with Dave on Apollo and the Space Shuttle program and he was a very aggressive manager. However, I Googled Harrison Storms several times and read the names of his Storm Troopers, and Dave was not one of them. The director reported to Gary Osbon, who was Apollo chief engineer, and that is the direction the pressure would follow down to us working troops. Now why would there be so much pressure coming to us just to get us back on schedule? Well, the schedule was not the only consideration. There were, by contract, dollar incentives to be paid by NASA to North

American Aviation if the AFRM 009 project could meet the contractual schedule and even higher incentives if the project could be completed earlier. These were day-by-day incentives that became bigger with each added day that AFRM 009 was delivered to NASA ahead of the contract date. I am talking here about millions of dollars. There were also heavy penalties if North American Aviation delivered AFRM 009 beyond the contracted schedule date, with greater penalties than the incentives given for beating the agreed-on schedule.

At this time in the Apollo program, AFRM 009 and all its systems (with the exception of the Automated System) were already three years into the design and manufacturing phases but still had many challenges to overcome before they could meet the NASA delivery schedule, let alone beat it. However, because the Automated System was in its infancy design stage, it had more work ahead just to catch up to AFRM 009 and its other systems. The Automated System thus had the highest risk of not meeting the NASA delivery schedule and jeopardizing the millions of incentive dollars. For this reason, the Automated System was referred to as the long pole in tent when anyone referred to the AFRM 009 delivery date. In 1964, AFRM 009 was the hot seat of Apollo because of schedule concerns, and that made me nervous.

2

AFRM 009 and the Eagle Knight Lead Engineer

In the meantime, in 1964, the main activity that was taking place in Building 6 in Downey in Bill Fouts' organization was that Bill was holding meetings with Carl Conrad, Larry Hogan, Bill Paxton, Tom Shanahan and other members of the AFRM 009 team to discuss the AFRM 009 mission and the design of the Mechanical Boy. Bill was also attending meetings with some of the NASA managers at the Manned Spacecraft Center in Houston to better define the AFRM 009 mission timeline. In the meetings in Houston, Bill gave NASA the status of the design of Mechanical Boy, which was then a design on paper only.

There were printed copies of a preliminary Procurement Specification Document for the Mechanical Boy to which Carl Conrad and Larry Hogan had contributed heavily. The document had been given to Autonetics, a Missile Division sister division in Anaheim (the same division that had mechanized the Hound Dog Cruise Missile Flight Control System). The Procurement Specification had been given to them under a work order called an Interdivisional Work Authorization to mechanize, build and test the Mechanical Boy. I went to document control to get a personal copy of the specification document because it was a most important piece of literature, as it had been on the Hound Dog cruise missile program. In the earlier program, my mentor Ed Kelley had said to me regarding the Hound Dog Cruise Missile Fight Control System specification, "Not only do you have to memorize it, but you also have to breathe it, sleep with it, eat it and digest it and live with it"—words that would serve me well on the AFRM 009 project and, for that matter, for the rest of my career.

Some background on Autonetics as to why it was considered the Missile Division's sister division and why it was selected to mechanize the Flight Control System for the Hound Dog cruise missile and the Mechanical Boy for AFRM 009: Around 1947, the U.S. government was researching the best method for building a defense system that could carry and launch nuclear weapons (namely, atomic bombs) when defending itself against an enemy. There were two competitive methods being considered, one being intercontinental ballistic missiles and the second the cruise missiles. Contracts were awarded for both types of weapons, and North American Aviation was awarded the Navaho program, which consisted of a cruise missile named the X-10 that was carried aloft by the Navaho booster rocket. The North American team that worked on the Navaho program consisted of the engineers who designed the booster and the X-10, those who designed the engines, and those who designed the flight control system for the X-10. The Navaho program was cancelled in 1957 because the United States decided to go with intercontinental ballistic missiles. However, once an intercontinental ballistic missile has been launched, it cannot be called back, and the government worried that in the event that the United States launched an ICBM in retaliation for an enemy's nuclear attack, and it was later discovered that it was a false alarm, the ICBM could not be stopped. The solution to the problem was to have both the ICBM and a cruise missile nuclear arsenal with cruise missiles, one under each wing of a B-52 bomber, being always ready for a first retaliation strike against an enemy's nuclear strike. If it turned out to be a false alarm, the B-52 with its two cruise missiles could always be called back. So later, around 1960, the Air Force at Wright Patterson Air Force Base awarded North American Aviation a contract for the Hound Dog cruise missile based on the X-10 design. It was at that time that North American Aviation formed the Missile Division with the section of the engineers who designed the booster and the X-10, along with a separate division named Autonetics with the engineers who designed the flight control system. The Missile Division was the prime contractor for the Hound Dog cruise missile, and because the Autonetics sister division had experience in designing the flight control system for the X-10 using solid-state devices, the Missile Division awarded the Hound Dog Cruise Missile Flight Control System contract to Autonetics. When the Missile Division received the Apollo AFRM 009 contract, the Missile Division also awarded the control programmer contract to Autonetics because they had abundance experience in designing systems with solid-state devices, which was needed for space flight rather than vacuum tubes.[1]

The story of Autonetics and some of the Navaho program history that I have written about here was told to me by my boss and mentor Ed Kelley when I started working as a research engineer on the Hound Dog cruise missile at the Missile Division of North American Aviation. One day in 1960 when I arrived at work from my last class at USC, Ed Kelley came to me and said, "De La Rosa, get your stuff together; we are going to Autonetics." All the way to Autonetics, Ed told me the history of Autonetics, and on the way back to our home plant Ed told me more history of the Navaho program. Ed was a World War II veteran and worked on the Navaho program X-10 cruise missile. It wasn't until 1995 that I started writing my first manuscript about my experience during 1960 to 1964 in helping design the flight control system for the Hound Dog cruise missile, which I completed in 2000.[2]

The objectives of the AFRM 009 mission were to (1) test the Thermal Protection System (the heat shield) under outer space and reentry heating conditions that the capsule would experience on its return trip from Earth orbit, (2) test the Service Module Propulsion System operation, (3) test separation of the boosters and the Command Module–Service Module combination, (4) test the Reaction Control System operation at altitude, (5) test the stability of the boosters, (6) test the stability of the Command Module and Service Module when connected, (7) test normal Launch Escape Tower jettison, (8) test the Environmental Control System operation, (9) test the Electrical Power System operation, (10) test the fuel cell operation, (11) test the parachute system that would ensure a safe landing for the astronauts in the ocean after their return trip from outer space, (12) test the cushioning system used for the astronaut seats that would protect the astronauts from the landing impact on the water, and (13) test the capsule recovery system and recovery techniques. The flight would also serve to test the capsule's airframe structure under Earth reentry and water landing conditions and stresses created by the launch environment, space travel, and reentry and landing environment. These were all important and critical milestones because the first scheduled Apollo manned mission was to be an Earth orbit mission in preparation for Apollo deep space missions.

On my fourth day of working on the unmanned Apollo capsules, I began to make phone calls to engineers in Building 6 and Autonetics engineers in Anaheim for appointments to go see them to discuss the AFRM 009 mission and Mechanical Boy design. One of the first courses of action, of course, was to find out who the Mechanical Boy players were at Autonetics. Autonetics had a small team working on Mechanical Boy, and the project engineer was Bob Wark, the same person I had worked with on the flight control system for the Hound Dog cruise missile program when Autonetics was the subcontractor that mechanized the system for the cruise missile. With the name of the project engineer of the Mechanical Boy Automated System at Autonetics, work on the system design could begin. On a telephone call with Bob Wark regarding design work on Mechanical Boy, Bob invited me to come to Autonetics and meet with him and whoever else was working on the project. Bob requested that I come over that same day and he would have a group of people waiting. According to Bob, they had been anxiously waiting for someone from the Downey division to come over to Autonetics and give them some direction. After getting directions for how to get to Autonetics and the location of Bob's building, I immediately got on the road. When I arrived at Autonetics, Bob was in the lobby waiting for me and proceeded to get me a visitor's badge, and thereafter we walked into the engineering area.

While Bob and I were in the engineering area, the Autonetics engineers who were at the meeting were Al Okamura, whom I knew from the Hound Dog cruise missile program; John Rowe, the Mechanical Boy lead designer; and Sam Rotuna and Vaughn Slaughter from the Project Office. Another project engineer from Autonetics who was on the project was Jerry Paccassi, who was also on the previous Hound Dog Missile program. Bob Wark got his team together in a conference room, and the discussion centered on the Mechanical Boy design, at which time many questions were asked. By the end of the discussion there were plenty of action items to work on, areas where Autonetics was looking for direction.

Back in Downey that same day, I worked the action items brought back from Autonetics. The next day was Friday, and at work that morning I had completed all the action items brought back from Autonetics. As I was getting ready to call Bob Wark, I started talking to a new engineer who had taken one of the empty desks. The new person said his name was Thom Brown and that he had transferred in from Field Engineering. As we were talking, I told Thom that I was working on some action items from Autonetics received the previous day. Thom seemed interested in what I was telling him, so I told him that Autonetics, our sister division in Anaheim, had been selected to build into hardware and test the Mechanical Boy that had been designed here at our division. I told Thom that in engineering we refer to their task as mechanization of our design. I further told Thom that I had been to Autonetics and had talked to them about the mechanization they were pursuing based on the Mechanical Boy design in the Procurement Specification and that the action items I had worked on were those I had received from Autonetics. Thom knew that none of us had received any kind of assignments, so he asked, "How in the hell do you know what to do?" The answer I gave Thom was "I have worked R&D programs before. As a matter of fact, I just transferred in from an R&D program. When you've worked on R&D programs you learn what it is that has to be done." Thom's face had a look of a man saying to himself, "Here is a guy that is working and knows what to do, and I cannot do anything because I do not have an assignment yet."

Because he was showing interest, I told Thom, "I'm going back to Autonetics today; would you like to go with me?" Thom jumped at the chance and said, "Yes, I'd like to go." After talking to Thom, I called Bob Wark and told him that Thom and I were coming over. I put all my material together, and Thom and I headed out the building to the parking lot and drove the twenty minutes to Autonetics to meet with Bob. As we walked into the engineering area, I introduced Thom to everyone in the Autonetics team. My hope was that Thom would become one of the important players on Mechanical Boy Automated System for AFRM 009.

During the following week in Downey, Thom and I kept working with the Autonetics engineers, and Bill Fouts finally documented how he was going to organize his department. There would be two groups to do the Automated System task. One group would be the Automated System Requirements group headed by Larry Hogan, and the second group was the Automated System Design group headed by Bill Paxton. Those engineers who were already occupying desks in the area were assigned to one or the other of the two groups. Thom and I were assigned to work for Bill Paxton in the Automated Systems Design group, and Carl Conrad was assigned to work for Larry Hogan. Members of both groups were moved to areas and desks that had been assigned to each of the two groups.

As soon as both design units were settled, Bill Paxton called me into his office, and as soon as I entered, he said, "I would like you to be my lead engineer." Of course, I agreed. I sat down in one of the chairs, and we talked of what lay ahead for the design unit. We had one big job to do, and my thought was "If Paxton and I can work together as Ed Kelley and I did on the Hound Dog cruise missile, we will have no problem doing the job." When we were through talking, I walked out of the office thinking that now we would see some action. The lead engineer of any project is responsible for assigning the required project work to his team members, as opposed to the project or program responsibilities of management. This means that any engineering action that takes place on the project is orchestrated or directed by the lead engineer, so I was very aware of what lay ahead for me because of my previous association with the Hound Dog cruise missile program.

A day later, the moving people brought me a conference table to go with my desk; they also brought one for Carl Conrad, who was now Larry Hogan's lead engineer. While I was rearranging my desk and conference table, I looked up and down the hall and saw the chief engineer, Gary Osbon (who had also been the chief engineer on the Hound Dog cruise missile), walking down the main aisle in my direction. As Gary got closer to our design unit, he looked my way and spotted me. He turned and came down the aisle, walked over to me, and as he shook my hand, he said, "Welcome to Apollo, Jim." He asked how things were going and whether everything was okay. I told him, "I'm doing fine and just getting settled." Then he smiled and said, "You know that here we have no jitter!" (He was referring to a major problem with the Hound Dog cruise missile that I had helped solve; if it had gone unsolved, it could have resulted in contract cancellation by the Air Force. Salvaging the contract had resulted in seven hundred cruise missiles being delivered to the Air Force by North American Aviation Space Division.) He shook hands with me again, and as he was leaving, he said, "Have a good time." There is no recollection in my mind of having contact with Gary again during the time I was on the Apollo program, but it was nice of him to come welcome me to Apollo. The chief engineer of Apollo traveled in higher circles than most of us workers, and so there was no need for us to meet with him.

In the April 17, 1964, issue of *Skywriter* (North American Aviation's company newspaper), it was announced that John Paup had been promoted to vice president and assistant to Harrison Storms, division president, and Dale D. Myers had been promoted to vice president and Apollo program manager. Dale had been program manager of the Hound Dog cruise missile program with Sandy Falbaum as his assistant. It was also around this time that the Space & Information System Division *Skywriter* (in the same issue as listed above) announced that engineers had developed the Apollo center probe and drogue rendezvous technique (different from the original design) specifically to join the Apollo capsule with the lunar landing module. The center probe and drogue rendezvous technique was originally developed during a flight of Gemini 6 and Gemini 7 in a combined fashion like the one used for inflight refueling of military aircraft using a conical funnel apparatus.[3]

So, in April 1964 the Apollo program was rapidly moving forward. By the end of April, the Unmanned Apollo Automated Systems group was already functioning as a design group. In addition, R.C. Kellett, the project engineer on Mechanical Boy, had issued an internal letter saying that the name "Mechanical Boy" was no longer acceptable to NASA and that the accepted name was "Control Programmer." From that day forward, the name "Mechanical Boy" disappeared, and Control Programmer became the official name of the automated system for AFRM 009. Because Control Programmer was too long to use in everyday conversations and internal letters, everyone began to call the system *CP*. Likewise, the name "Mechanical Man" for the automated system for the other four unmanned Apollo spacecraft was changed to "Mission Control Programmer."

For the reader to understand the AFRM 009 mission and how the Control Programmer controlled AFRM 009 during its mission, a discussion of the Control Programmer is in order. But first the mission of AFRM 009 must be understood: AFRM 009 was to be launched with the Saturn I booster as the first stage and the Saturn IVB as the second stage. There was to be no Lunar Excursion Module (LEM) in the rocket stack. The Service Module would be attached to the Saturn IVB booster with the Command Module connected as a unit to the Service Module. The Launch Escape Tower was at the top of the stack, attached to the apex of the Command Module.

During a moon mission, the function of the Service Module was to accompany the Command Module and the Lunar Excursion Module to the moon and back. On the trip toward the moon, the Command Module would serve as the habitat for the three astronauts and contained all the systems for control of the lunar traveler on its way to the moon. The Lunar Excursion Module was to be used only for landing on the moon and for blasting off the lunar surface and then reconnecting with the Command Module–Service Module combination that was parked in moon orbit. Once the LEM had completed its mission, it would be jettisoned into space to orbit the sun on Apollo's return trip back to Earth. The function of the Service Module on the lunar and return trips was to provide the needed propulsion to make course corrections during both trips and to provide the propulsion needed for going into moon orbit and later for slowing down the Command Module for the return to Earth. The Service Module also contained the fuel cells that provided electric power for the long moon journey, in addition to providing the crew with potable drinking water that was a byproduct of producing the electric power. The Command Module had batteries that had only enough power for the capsule's Earth reentry period, meant to supply power to all systems required for a safe landing on the ocean surface. The Service Module contained one rocket engine officially named the Service Propulsion System engine (which we always called the SPS engine). This was the engine that was to do all the heavy work on the travel to the moon and on the return trip. The Service Propulsion System engine was an extremely critical piece of the Apollo program and was required to be ignited, turned off and then reignited again several times. Igniting and then reigniting a rocket engine of this type had never been accomplished in outer space before. If the mission to the moon was to be successful, this feature of the Service Propulsion System engine had to be proven in space before a full-blown manned moon mission was attempted. The Service Module also contained all the fuel tanks required by the Service Propulsion System engine. The Service Propulsion System engine had a nozzle that was gimbaled (meaning that the engine nozzle was movable in any direction to position the engine thrust in the desired direction for any specific desired maneuver required of the Command Module–Service Module combination). The control of the Service Propulsion System engine was executed by the astronauts or automatically by the computer in the Command Module. The Command Module contained all the control panels with the necessary switches for commanding everything that had to happen for a successful lunar mission.

Because AFRM 009 was the very first flyable Apollo Command Module to go into space, it was unmanned, and the Control Programmer was to perform all the functions normally performed by the astronauts if they had been there to execute the mission. It was the function of the Control Programmer to orchestrate the graceful movements of AFRM 009 in space when propelled on its history-making voyage to usher in the flight of the Apollo program. The AFRM 009 Control Programmer was to be made up of two major functions, one being the automatic function and the other being the radio command function. For this reason, the Control Programmer was made up of two electronic boxes: the Automatic Command Controller and the Radio Command Controller.

So why did the Control Programmer require two functions? For the AFRM 009 mission, the Control Programmer's function was to replace the astronauts. Also, by NASA's directive, the ground rule for the AFRM 009 mission was that no single failure of the Control Programmer should cause the loss of the Command Module. There were many possible failures of the Apollo boosters and modules that could result in explosions or cause the Command Module to burn up on reentry into the Earth's atmosphere. Because of NASA's directive, Autonetics had to ensure that all Control Programmer failures of automatic functions that could occur by electronic circuit failures be made redundant so that the electronic failure would not cause loss of the Command Module and the function could still be executed even with the one electronic fault. If this task could not be accomplished with redundancy of electronic components, then the failed function should be made to fail in the benign state so that it could be accomplished by radio control from the ground. The function of the Radio Command Controller was to contain all these radio-controlled functions and to facilitate its communication with the ground radio controller when required. There were Apollo failures that could occur early during the launch phase of a mission that could cause a mission abort whereby, in an Apollo mission with three astronaut crewmen on board, it was planned that during the early mission abort after the Apollo Command Module separated from the booster, the astronauts would take control of the Command Module and, with the Rotation Hand Controller, issue commands to position the Command Module for a safe Earth landing by visually looking out the window. Since the Control Programmer was flying the Command Module blindly and could not see out the window, it could not handle these types of aborts, so it was planned that on AFRM 009 these aborts would be handled by a controller on the ground in communication with the Radio Command Controller.

To perform its normal functions during its mission, the Control Programmer would have to perform the function that a switch on the Command Module control panels would perform when moved by an astronaut. The Control Programmer also would have to interface with the system that this switch commanded to perform a function. First, the Control Programmer would have to jettison the Launch Escape Tower. It would then have to uncage the Stabilization and Control System gyros (the system that Paul Garcia and Danny Moreno worked on), which would be used to maneuver the Command Module to the correct orientation for Earth reentry. The Control Programmer would subsequently have to command the separation of the Command Module–Service Module combination from the Saturn IVB booster. The Control Programmer would have to command the Service Propulsion System engine to ignite and turn it off at the proper time, and then reignite the engine and again turn it off at the proper time so that the Command Module would return to Earth at the correct point so that it would land at the place in the South Atlantic Ocean where the Navy ships were waiting to recover it.

The AFRM 009 mission, at first glance, appears to be easy, but it was an extremely difficult task. To begin with, at the time, no one had ever put a real flyable Apollo capsule into space, much less into orbit around the Earth. The systems on the complete Apollo rocket had to work perfectly together, and all North American Aviation had at the time to ensure that this could happen was a concept and a computer simulation that predicted that if all the mission assumptions were correct, the AFRM 009 would have a successful mission. In addition, the original AFRM 009 mission was redefined by a new mission, and the status of the Control Programmer design was briefed to NASA in a meeting with Bill Fouts, R.C. Kellett and Tom Shanahan of the Mission Design group at North American Aviation on May 26, 1964.

The automatic functions that the Control Programmer would have to perform were a timed sequence of events during the launch flight phase starting from the separation of the Saturn IVB booster and the Command Module–Service Module assembly in sequence all the way to reentering the Earth's atmosphere, including ocean splashdown. The mission was to be performed without a moon-mission-type guidance and navigation system on the capsule, and all the functions performed by the Control Programmer were timed events. The method that was conceived at North American Aviation by Carl Conrad for performing the Control Programmer timed events was using a timer run by an electric motor. This timer was an off-the-shelf device, which meant that it needed no design or development time. The design and development of the timer had all been done previously by Lockheed in Sunnyvale near San Jose, California, for the Atlas-Agena booster. Carl knew of the existence of this timer because he had worked for Lockheed on the Agena program while he was pursuing his physics degree at the state university in San Jose. Carl obviously knew some people at Lockheed who provided him with all the engineering information for the Downey division to issue a Specification Control Drawing to Autonetics so they could issue an order to Lockheed for the purchase of Lockheed timers that suited the needs of the Control Programmer and in sufficient quantities to support the AFRM 009 program. The Lockheed timer motor would operate the timer that would close or open switches at the specific times required by the AFRM 009 mission. Many months later, Bob Antletz would joke with Thom Brown that he was working on a washing machine timer. Bob would ask Thom, "Is the washing machine timer in the wash cycle or is it in the spin cycle"? Then Bob would break out into a laugh. Bob may have been razzing Thom, but that is exactly the way the timer worked—like a washing machine timer, only at a faster pace.

The things that had been discussed with Bob Wark at Autonetics prior to the department being organized concerned the mechanization of the Control Programmer as defined in the original AFRM 009 mission. Now, however, the mission had been modified by North American Aviation and NASA, but Autonetics had not yet been informed about it in detail.

There were also some other loose ends that had to be taken care of before Autonetics could really get to work on the Control Programmer. Up to this point the Control Programmer mechanization of the North American Aviation Missile Division design that had been reviewed and discussed by me and Thom Brown with Autonetics was a

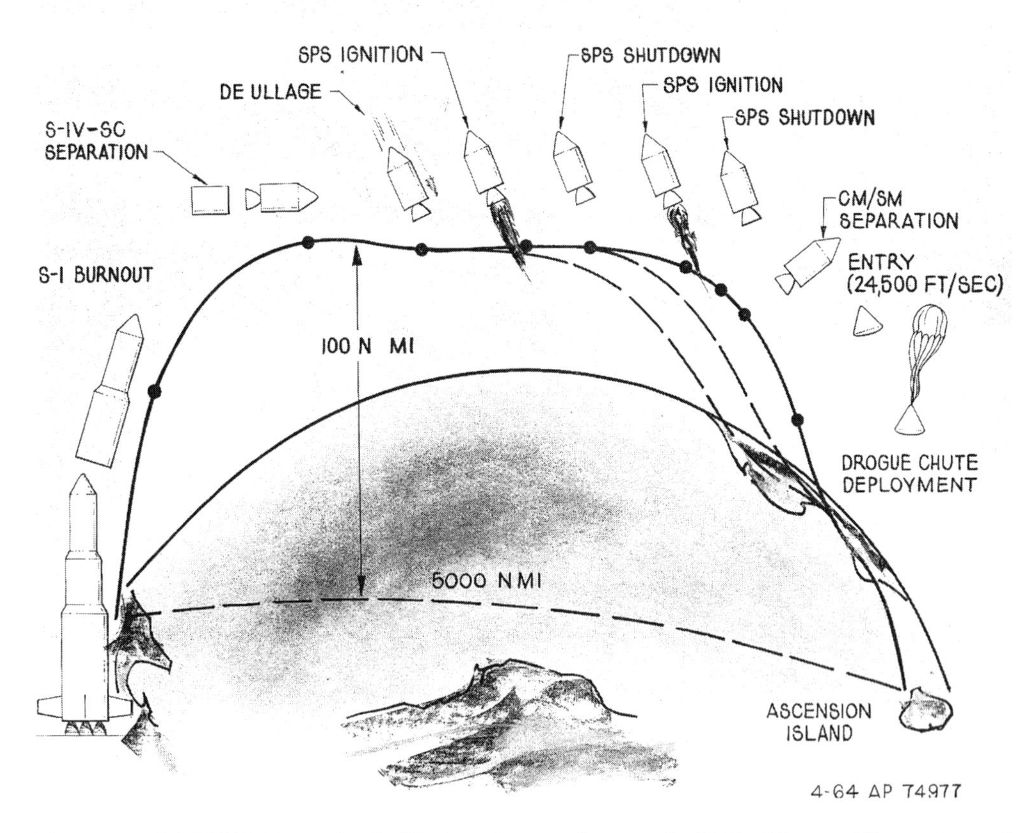

The original AFRM 009 mission ballistic trajectory from early 1964 without rolling entry and without high Earth orbit reentry velocity (Courtesy of the Boeing Company).

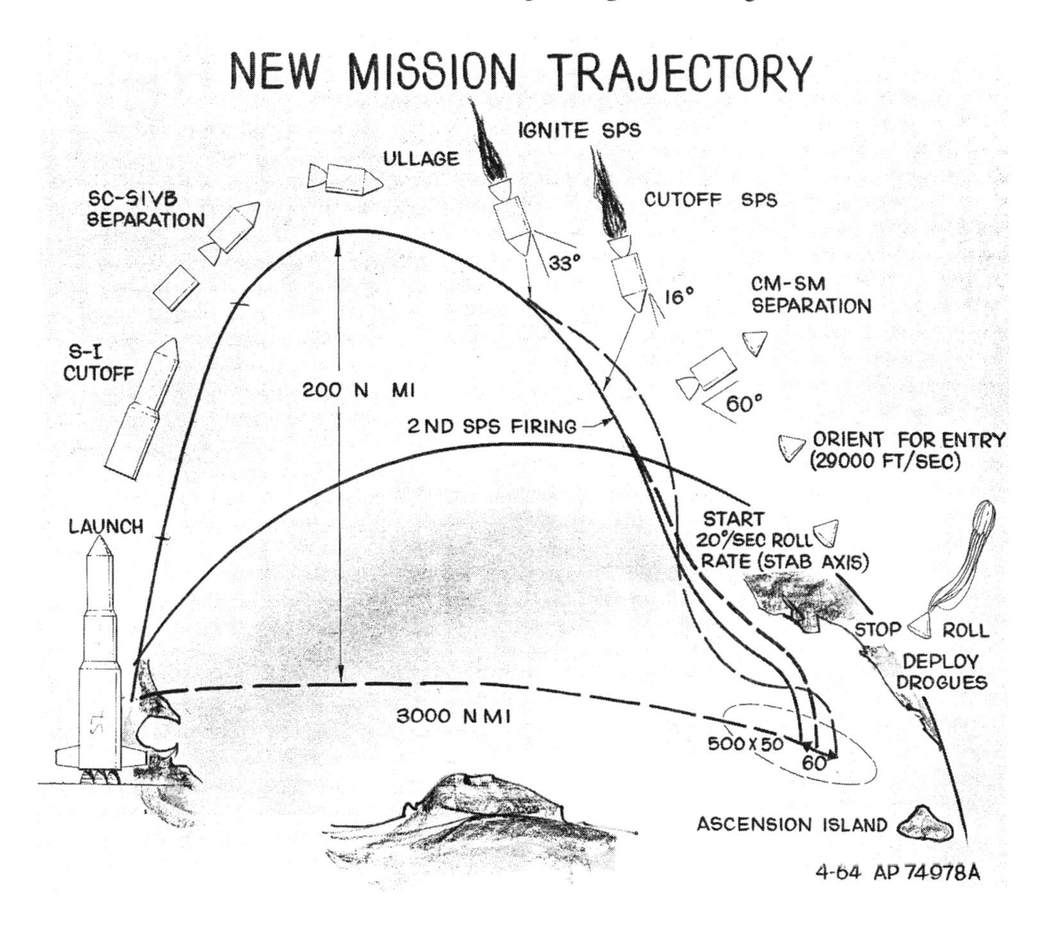

The new AFRM 009 mission ballistic trajectory from late 1964 with added rolling entry and high Earth orbit reentry velocity (Courtesy of the Boeing Company).

design on paper only. The mechanization that Autonetics had been putting together was documented by John Rowe in a system schematic diagram that Autonetics believed was consistent with the design given to them in the Control Programmer Procurement Specification. There are other documents that are important in a system's design and necessary for purchasing a system from another company or from a division within your own company. What follows is a description of these other documents: First, because the missile division had the Apollo contract directly from NASA, it was known as the prime contractor. This allowed NASA to send specialists to the facility in Downey or have the NASA specialists in residence at the Downey division to monitor or redirect the division on the Apollo program, providing that those resident specialists had been given the proper authority from either Washington or one of NASA's major centers. In addition, NASA had a subsystem manager who worked out of the Manned Spacecraft Center outside of Houston, Texas, for each of the Apollo systems who visited periodically to meet with North American Aviation and subcontractor personnel whenever necessary. The NASA subsystem manager for the Apollo Automated Systems was Gene Holloway, who met with Downey engineers on a periodic basis.

The North American Aviation Missile Division was in the business of designing satellites, missiles, and spacecraft. In only a limited way did those at North American

NEW MISSION
FUNCTIONAL REQUIREMENTS

• NORMAL SEQUENCE

EVENT		TIME (SEC)
ELS "SAFE" COMMAND		165 SEC FROM LAUNCH
REMOVE ELS "SAFE" COMMAND		166 SEC FROM LAUNCH
UNCAGE GYROS	0	604 SEC FROM LAUNCH
S-IVB-SC SEPARATION COMMAND	0	
SCS ΔV MODE ENABLE	5	
PROVIDE TM CALIBRATION SIGNAL FOR 10 SECS	5	
S-IVB-SC SEPATION COMMAND OFF	6	
COMPLETE PITCH MANEUVER	31	$\Delta\theta \approx -62°$
ULLAGE ON	295	
IGNITE SPS	315	
ULLAGE OFF	316	
SPS OFF	550	

4-64 AP 75068

According to this sequence, T-zero for the Control Programmer would occur at 604 seconds after launch, at the same time that the BMAGs in Block I Stabilization and Control System were uncaged. After "S-IVB—SC Separation Command," the first CP event would be "SCS ΔV Mode Enable" at 5 seconds elapsed time after t-zero (Courtesy of the Boeing Company).

Aviation have the capability or capacity to build the electronic systems that went into the craft that they designed and built. For this reason, they depended on outside companies to mechanize and build the electronic systems needed on the Apollo program. When contracted by the Downey division to do the system work, these companies were considered subcontractors. Both NASA and North American Aviation by necessity had specialists communicate and conduct business with the subcontractors and visit the subcontractor facility at will, providing they had been given prior authorization. If the subcontractor was not within the vicinity of the Downey facility, then the division would have resident specialists at the subcontractor facility. For some systems, the North American Aviation specialists had moved with their families to be the division's resident representatives to whatever town or city where the subcontractor was located. Fortunately for those of us working on the Automated System for AFRM 009, Autonetics was in Anaheim in Orange County, and no one had to uproot their families and move out of state.

The contract for the AFRM 009 Control Programmer was awarded by the Downey division to Autonetics in early 1964 prior to my arrival to the Apollo program. The contract was awarded by issuing a minor Interdivisional Work Authorization so Autonetics could do the preliminary work on the Control Programmer. Autonetics was also given the Control Programmer Procurement Specification. These two documents were called the Procurement Spec and described the CP design and the minor IDWA that provided the money to do the work, which in essence was the statement that spelled out what work was to be done, what product was to be delivered to the Downey division and the schedule for when that product was to be delivered. A third document was also necessary: the Specification Control Drawing (SCD). The Specification Control Drawing defined the envelope of the product and other physical characteristics. Autonetics was given two copies of the Specification Control Drawing—one for the Control Programmer and one for the Lockheed timer.

What I have given here is some of the Apollo jargon that was used during the program. To avoid confusion, I have minimized the use of such jargon because if I used it as extensively as it was used during the Apollo program, reading all that I have written would be impossible without having a list of acronyms to use for translation. The jargon

NEW MISSION

FUNCTIONAL REQUIREMENTS (CONTINUED)

- ABORTS
  - LES ABORT, SAME AS ORIGINAL MISSION
  - NO PROGRAMMER REQMT FOR ABORT FROM BOOSTER AFTER LES JETTISON
  - AUTOMATIC CONTROL IF SPS FAILS TO IGNITE
- GROUND CONTROL BACKUP
  - CORRECT FOR SINGLE G & C FAILURE
- POWER SYSTEM FAILURE PROTECTION
  - DISCONNECT FAILED FUEL CELL
  - REPLACE FAILED INVERTER
- ELECTRONIC MISSION SUCCESS OF .984
- TIMED ACCURACY - 1%

4-64 AP 74979

The requirements for AFRM 009 Mission Abort, ground control for failures, power system failure protection, a probability of success of 0.984 and a timed accuracy of 1 percent (Courtesy of the Boeing Company).

came from almost any organization that was associated with the Apollo program—from NASA in Washington, from all the numerous NASA centers, from North American Aviation and its hundreds of subcontractors. The NASA acronyms can be found online.

It was imperative that the Control Programmer design changes required by the new mission and any other change to the Control Programmer be given to Autonetics as soon as possible because even though, as mentioned before, the Autonetics Control Programmer mechanization was on paper only, there were several facets of any system procured by the prime contractor from a subcontractor for the Apollo program that, if not attended to quickly, could end up jeopardizing the timely delivery of the product. The subcontractor had to hurry and complete the system mechanization as fast as possible and have it approved by the prime contractor so that electronic, electromechanical, and mechanical parts for their system could be ordered from suppliers of these components for the subcontractor to meet their system delivery schedule and because all these parts for the Apollo program were seldom bought off the shelf. Most electromechanical and mechanical parts for systems were specially made for the Apollo program and had long lead delivery times. In addition, electronic parts on systems for Apollo were required to be highly reliable electronic components that had survived months of rigorous environmental testing. The electronic components also required a high pedigree trace ability for each individual component and, for that reason, were unbelievably expensive.

The need for high-reliability parts stemmed from the need to ensure that an electronic component would survive an Apollo mission without a failure that would cause loss of human life or, in the case of an unmanned flight, would not cause the loss of an expensive rocket. The ordering of parts for the Control Programmer redesign to the new AFRM 009 mission was not the only problem that Autonetics had. There was also the added cost and the change in product delivery schedule that would result from the design changes to the Control Programmer. In addition, before Autonetics could do anything, the Downey division had to give them written direction on what changes had to be made to the Control Programmer to bring it in line with the new AFRM 009 mission, and it had to be done in a formal fashion. The Automated Systems Design group at North American Aviation would first have to revise the Control Programmer Procurement Spec to reflect all the changes to the Control Programmer resulting from the new mission. The design group would also have to issue a new IDWA to replace the minor IDWA that Autonetics had received to do the preliminary Control Programmer work. But even before all this new work could be started, the present Control Programmer mechanization based on the original Control Programmer Procurement Specification had to be agreed on by the Downey division and Autonetics as a baseline.

The very first thing that had to be done was to communicate to Autonetics the results of the Control Programmer system review that had been made by me and Thom Brown prior to Bill Fouts' department being organized, against the current released version of the Procurement Specification. This had to be done, as mentioned before, so that an agreed-to baseline between Autonetics and the Downey division would exist so that future design changes could be made by the Downey division and transmitted formally to Autonetics. Autonetics also had to give the Downey division a firm cost on the agreed-to baseline so that any cost resulting from future design changes could be provided by Autonetics to the Downey division. The things that had been identified for change to bring the Control Programmer mechanization in line with the Procurement Specification were to reduce or delete some of the Control Programmer redundancy.

The inclusion of four Apollo capsules to fly unmanned missions added cost to the already expensive Apollo program. There were other aspects that made the Apollo program an expensive endeavor besides the need for high-reliability electronic components and the addition of four unmanned Apollo missions. One of the main issues that drove the cost so high was that the Apollo program was a throwaway program—that is, every Apollo capsule shot into space, whether it was to perform a test, to fly an unmanned capsule or to go to the moon, was put into space by expensive rocket boosters that were not retrieved to be reused but were dropped back toward Earth to either be destroyed when they crashed or burn up during reentry. On both unmanned and manned missions, the only part of the Apollo rocket that returned to Earth was the Apollo capsule. On moon landing missions, even the moon landers were thrown away, and the vehicles used to drive on the moon on some missions were left behind. None of the Apollo capsules were ever reused for succeeding flights; every mission used a new capsule. It was not purposely planned that way, but of all the methods that were proposed for getting to the moon, this was the only method that had a chance of putting Americans on the moon in time to meet President Kennedy's timetable of reaching the moon by the end of the decade.

The clock to the end of the decade was ticking, and no one was sure, given all the work that remained to be done, whether the schedule to support a moonshot in 1969 could be met.

3

The AFRM 009 Control Programmer Team and a New AFRM 009 Mission

Sometime around the end of May or the beginning of June 1964, Bill Paxton assigned two engineers to my team for the design of the Control Programmer: Gene Rucker, whom I knew from the Hound Dog cruise missile days, and Don Farmer, who had recently been hired. A little later in the year, Z. Stanley Kleczko (who had been on the Apollo program longer than anyone else on the team) was assigned to my team to handle the Attitude Reference System that had been given to us in addition to the Control Programmer. The Attitude Reference System was an off-the-shelf system of attitude gyros that North American Aviation was purchasing for AFRM 009 and was a backup to the Stabilization and Control System Attitude Reference System. The Attitude Reference System also displayed the Command Module attitude on a visual display to the ground controller who was assigned to handle the early launch aborts that the CP could not manage; in this way, the Radio Command Controller on the ground could land AFRM 009 safely after a launch abort. Because Stanley had been working on Apollo longer than anyone in our team, he was a great help. When there were problems that required help from other Apollo groups, Stanley invariably could tell me who to go to for help and would in most cases know the man personally and could call them to make appointments to meet and talk about potential solutions.

Starting in June 1964, the year was full of activity for those of us who were working on AFRM 009, and it continued all the way to the end of the year. The main objective was to define a Control Programmer design that Autonetics could mechanize for AFRM 009 and that would successfully execute the timeline for the AFRM 009 original mission. This mechanization for the original AFRM 009 mission, as mentioned previously, would serve as a baseline mechanization from which changes to the new AFRM 009 mission could be identified and serve as a program dollar cost baseline from which Autonetics could add or subtract dollars for design changes that the missile division would make to the Control Programmer. It was my responsibility to know everything about the Control Programmer, how it worked and the function of each component in the Autonetics schematics, and what part each component played in the operation of the Control Programmer; this was not difficult for me because of my experience teaching radar systems electronics in the Air Force, as well as my experience with the design of Airborne Radar Systems test equipment at Hughes Aircraft Company and, of course, the Hound Dog cruise missile design. A physical Control

Programmer in the form of hardware would not or could not be available until sometime in 1965.

The question that had to be answered at work was this: How does a change from the original mission to a new mission affect the design of the Control Programmer? For the answer to that question, let us look at how the Apollo Command Module would have behaved during a mission (the explanation that follows is lengthy, so please do not blink): The AFRM 009 Command Module was to be carried aloft by two large boosters—the Saturn I and the Saturn IVB. The Saturn I booster engines were to be ignited on the launch pad. The force of the Saturn I booster engines would then carry the large rocket assembly to a defined altitude where the Saturn I booster engines would have imparted a sufficient force to the whole assembly to achieve a predetermined velocity, at which time the engines would be cut off by an automatic issued command. The velocity that the Saturn I booster engines imparted to the launched rocket, however, would not be sufficient to get the Command Module into space, so the Saturn IVB J-2 engine would then be ignited to give the assembly the required velocity to put the Command Module into space of zero gravity. The velocity required was the amount needed by the Command Module to overcome the pull of Earth's gravity. But before the Saturn IVB engine could be ignited, the Saturn I had to be separated from the Saturn IVB. This was accomplished by one of the systems on board the rocket by issuing a Saturn I Separate Command.

There were several methods of holding sections of rockets together so that they can be separated safely during the boost phase of flight. One method of holding the rocket sections together (which was the one used for the Apollo program) was to use explosive bolts. However, the bolts would not explode and blow apart during separation, because if that happened, the whole rocket would probably blow up. These were specially made, huge bolts that had a scarring around the diameter at the midpoint of the bolt length and contained explosive material in the inner core adjacent to the outer scarring. The explosive material was electronically ignited to break the bolt apart by excessive pressure at the scarring to allow the attached units to separate. Because all methods of separating rocket sections were explosive methods, they could not, for safety reasons, be in the least connected to an electric source. They were physically disconnected from any path that led to an electrical source from the time of lift-off to just prior to their detonation. The sequence for their detonation was to first connect the explosive device to a path that could lead to an electrical source (this is called arming the device) and then issue the command to ignite the explosives in the device to break the bolts apart and allow boosters' separation.

The other parts of the rocket that had to be dealt with were the bundles of electrical wires that connected one piece of the rocket to each other. These bundles of wires (or cables, as they were called) that transmitted electrical signals to the various systems also had to be severed before the two rocket sections could separate. These cables were cut by a guillotine method commonly referred to as dead-facing. When the physical separation of the Saturn I and the Saturn IVB was completed, the Saturn IVB engine was ignited to pull away from the Saturn I booster. The Saturn IVB continued to speed toward space, and the Saturn I would fall away to break apart and burn up as it entered Earth's atmosphere. When the fuel in the Saturn IVB had been exhausted, the procedure for separation was repeated for the Saturn IVB and the Command Module–Service Module combination. After the Saturn IVB separation, the Command Module–Service Module combination would go into a suborbital ballistic trajectory, meaning that, if left alone, they would fall back to Earth on their own, but not where the Navy ships would be waiting. Very possibly the landing site would be somewhere on the African continent. At this point, the SPS engine of the Service Module would come into play.

The function of the Service Module was to service the Command Module. The Apollo Command Module was relatively small and could carry only enough electrical power, oxygen, fuel, and other necessary consumables for the Earth reentry period. It was the duty of the Service Module to supply the Command Module during the longer periods of the mission. To position the Command Module for Earth reentry, the Service Module had some reaction control jets that were fired to achieve this result. Once the Command Module–Service Module combination was in position for reentry, the Service Propulsion System engine was then ignited to slow the Command Module or to speed it up, depending on what was required, so that it could reenter Earth in the position that would allow it to land in the vicinity of the waiting Navy ships. If more firings of the Service Propulsion System engine were needed to get the Command Module to reenter at the proper point, the Service Propulsion System engine could provide them. Once the Service Module had accomplished its function, it was separated from the Command Module by a Service Module Separation Command that would effect a separation of the Command Module and the Service Module as described previously.

In the AFRM 009 mission, the Stabilization and Control System would have to maintain the Command Module attitude for Earth reentry, which it did by firing reaction control jets contained on the Command Module. As the Command Module began to fall toward Earth, it would be weightless for a period, and then it reached a point where it was considered to start its reentry to Earth—namely, when it no longer was weightless and weighed a fraction of 1G, indicating that it was being captured by Earth's gravity. This reentry point happened at 0.05G. (1G is the gravity that humans, animals, and inanimate objects experience while standing on Earth.) There was a sensor on board the Command Module to sense the 0.05G gravity. After the 0.05G gravity point was reached, the Command Module reaction control jets were no longer significantly effective except for those in roll. The roll axis reaction jets allowed the Command Module to perform roll maneuvers only.

So, what is roll on the Command Module? If the Command Module is positioned with the blunt end down and the apex up so it looks like a teardrop and a pole is inserted into the Command Module at the center of the blunt end up toward the apex and then the Command Module is rotated on the pole, that rotation is roll and is called rotation around the roll axis. The roll of the Command Module was incredibly important in the reentry of the Command Module back to Earth because the amount and direction of roll provided to the Command Module by the Stabilization and Control System after it was captured by Earth's gravity could fine-tune the landing point of the Command Module as it impacted on the ocean surface and was accomplished by rolling the module to dissipate energy so as to slow down its velocity, thus enabling the deployment of the landing parachutes. When this roll maneuver of the Apollo capsule was performed during Earth reentry, it was called a "rolling entry." The rolling entry maneuver is not rolling the Apollo Command Module like a top but instead moving it in roll a few degrees in one direction and then reversing the roll in the opposite direction by the same amount. This is important information because the roll maneuver during reentry was not going to be tested prior to any Apollo manned missions prior to the decision of having an AFRM 009 mission.

How does the roll of the Command Module fine-tune the landing point and help dissipate energy? Well, for openers, if no roll is imparted to the capsule by the roll jets during Earth reentry, then the Command Module will come straight in, in a direct shot where it experiences the maximum lift. This reentry of the capsule with no roll maneuver is called a "lift vector up" reentry as opposed to the rolling entry. In a rolling entry, the Command Module will travel a zigzag path instead of a straight path, as in the "lift vector up" reentry. By following a zigzag path traveling the same distance, the module would splash down at a distance just shy of the straight shot because it would be falling at a steeper angle. In this way, by selecting the number of right and left turns and the duration of each turn, the landing point of the Apollo capsule could be pinpointed. The original AFRM 009 mission was a "lift vector up" entry, which was being changed to a rolling entry by the new mission. This certainly was a change for the Control Programmer.

The Service Module also had several sets of reaction control jets that were fired to position the Command Module for reentry, as well as for ullage maneuvers made prior to firing the SPS engine. The ullage maneuver was a plus-X translation of the Command Module–Service Module combination and was performed to move the Service Propulsion System engine fuel to the rear of the fuel tank and away from the engine combustion chamber to avoid explosions due to fuel that could have accumulated in the combustion chamber.

The heat shield of the Command Module covered the whole blunt end of the capsule and was ablative (meaning that some of the heat shield would burn off gradually during the reentry period). As the Command Module was exposed to the Earth's atmosphere, it met a great deal of resistance, which created friction on the blunt end of the Command Module. Friction creates heat, and the heat shield was very thick, so that the majority of the material would burn off as the Command Module reentered Earth. By the time the Command Module capsule splashed down on the ocean, the heat shield was black like charcoal and very thin. The ablative heat shield feature was also not going to be tested in space prior to the first manned Apollo mission, which became a major change to the Apollo program when AFRM 009 and all other unmanned airframe missions were added to the itinerary.

It was a standard procedure for a program as big as Apollo to have the various rocket systems interact with one another because, without the interaction, the mission would not have worked. Whenever two systems interacted with one another, we used to say that they interfaced with one another. The point where the electrical cables of the interacting systems met was called the interface. The Control Programmer of AFRM 009 interfaced with the Electrical Power System, the Ground Control Communications System, the Mission Event Sequencer Controller, and the Stabilization and Control System. The Control Programmer had to interface with the Electrical Power System because this was the source of the electrical power to power up all the Control Programmer electronic and electrical components. The interface with the Ground Control Communications System was necessary because, as was mentioned before, if some of the Control Programmer Automatic Command Controller functions failed, they had to be provided by the Control Programmer Radio Command Controller in communication with the Ground Control Communications System. The interface with the Mission

Event Sequencer Controller was critical because this system provided the Control Programmer with the times when certain rocket events occurred, which were used by the Control Programmer as cues for timing its own events, and the Control Programmer would cause some events to occur by commanding the Mission Event Sequencer Controller to execute them. The interface between the Control Programmer and the Stabilization and Control System is a bit more difficult to explain, so all that can be said is that when any movement of the Command Module was required, the Control Programmer accomplished this task by commanding the Stabilization and Control System to do what was necessary to make the desired Command Module movement by firing specific control jets.

Because the Apollo program was a manned crew program, the design of everything that was going to fly was designed with extensive safety factors. The Apollo flights had to be as safe as possible because human life was involved, national prestige was at stake, and expensive rockets had to survive through each flight. Each team of engineers that was designing parts of the rockets and systems to go on board any of them was careful

NEW MISSION

EVENT	TIME	
ULLAGE ON	550	
IGNITE SPS	565	
ULLAGE OFF	566	
SPS OFF	575	
COMPLETE PITCH MANEUVER	598	$\Delta\theta \approx +83^\circ$
PRESSURIZE CM RCS	598	
PRE-SEPARATION SIGNAL TO POWER PROGRAMMER	598	
ENTRY BATTERIES ON	598	
ENABLE SCS ENTRY MODE	598	
DISCONNECT FUEL CELL & SM BATTERIES	599	
SM-CM SEPARATION	604	
COMPLETE PITCH MANEUVER	635	$\Delta\theta \approx +65^\circ$
.05g SWITCHING	655	
BEGIN ROLL RATE	655	
ACTIVATE ELS	655	
ARM ELS PYRO	655	
PROVIDE TM CALIBRATION FOR 10 SECS	655	

4-64 AP 75069

At 655 seconds of the Control Programmer timeline, at .05g, AFRM 009 would be captured by Earth's gravity, and "Begin Roll Rate" (also at 655 seconds) would signal the start of rolling entry (Courtesy of the Boeing Company).

to include a degree of safety factor in the design. This is the reason why Autonetics had mechanized the original Control Programmer design with extensive amount of redundancy. The mechanization was based on the missile division's design, which was documented in the original release of the Control Programmer Procurement Specification. (The safety factor added to each system was what later saved the crew of Apollo 13.) The original AFRM 009 mission was designed to take AFRM 009 to an altitude of 100 nautical miles in a ballistic suborbital trajectory, requiring two Service Propulsion System engine firings to slow down the Command Module for reentry and travel 5,000 nautical miles down range to land near Ascension Island in the South Atlantic Ocean, just below the equator between Africa and South America, where capsule recovery would take place. AFRM 009 would spend nine minutes in zero Gs (weightless), the Command Module reentry velocity was going to be 16,704 miles an hour, and it would experience 16 negative Gs (deceleration). The reentry was a "lift vector up" reentry.

The new mission, however, was designed to take AFRM 009 to an altitude of 200 nautical miles, twice as high as the original mission, in a ballistic suborbital trajectory and 3,000 nautical miles down range to land in the ocean near Ascension Island. The higher altitude of the new mission was required because the new mission included

NEW MISSION

EVENT	TIME (SEC)
DEPLOY DROGUE CHUTE (25K FT SW)	975
TOUCHDOWN (IMPACT SWITCH)	1483
CONNECT POST LANDING BATTERY	1483
DISCONNECT ENTRY BATTERIES FROM MAIN BUSS	1483
CONNECT ENTRY BATTERIES TO POST LANDING BUSS	1483

4-64 AP 75070

The Control Programmer timeline commanded "deploy drogue chutes" before splashdown and ended with a command to "Connect Entry Batteries to Post Landing Buss" (Courtesy of the Boeing Company).

reversing the direction of the two firings of the Service Propulsion System engine in the original mission. Originally, the SPS engine was firing in a direction for slowing down the Command Module, but the new mission was firing the SPS engine in a direction to speed up the Command Module to expose the heat shield to a higher temperature. AFRM 009 would spend nine minutes in zero Gs (weightless), and the Command Module reentry velocity was going to be 19,283 miles an hour, which was faster than the original mission, and would experience 16 negative Gs (deceleration). The new reentry velocity was somewhat faster than the reentry velocity from Earth orbit and would certify the heat shield for the first manned Apollo mission, which was to be an Earth orbit mission. The reentry for the new mission was to be a rolling entry and would now require the Control Programmer to command the Stabilization and Control System to fire the capsule's roll jets for a rolling entry profile, which in the original mission was not required. The new mission also changed the timing of many of the mission events. But, as mentioned before, the Downey division could not authorize Autonetics to implement the new mission in the Control Programmer because the first task facing the AFRM 009 team was to get the Control Programmer baseline mechanization for the original mission before consideration could be given to sending the changes of the new mission to Autonetics.

The pressure was mounting on the AFRM 009 team, and the question was whether Bill Fouts' department and Autonetics could do all the necessary work on the AFRM 009 Control Programmer and still meet delivery schedules to support a 1969 moonshot.

4

The Control Programmer

For the AFRM 009 Control Programmer team, the year 1964 was a busy one. Meetings were the order of the day. At this time, in the Unmanned Apollo Program, meetings were necessary because the Automated System Design Unit had to organize itself, and the subcontractors liked a well-oiled machine. All parts of the AFRM 009 team had to stay exceedingly focused on the program because no Apollo flyable capsule had ever flown before and no one knew exactly what to expect during a mission to the moon (and especially what to expect during Earth reentry). Fortunately for the AFRM 009 team, Apollo was not designed in a vacuum. Every system designer knew what every other system designer was doing. There was an established communication link between all system designers that included letters between design groups, meetings between various engineers, and telephone conversations. Without this communication, the Apollo program and the landing on the moon would have never come to fruition.

All team members realized that much of the success of the Apollo program depended on the success of the unmanned missions. Everyone at Space and Information Systems Division (the new name of our division, sometimes referred to as S&ID) realized that the Apollo program had to succeed so that the United States could win the space race for national prestige; more important, Apollo had to succeed because astronaut lives were at stake. The largest product outputs during the year were paper documents such as internal letters within the Space and Information Systems Division that provided program direction, AFRM 009 system design direction, letters of direction to Autonetics, contractual documents to Autonetics and to American Wianko on the ARS system, engineering documents within S&ID and engineering documents to Autonetics. The most important meetings in 1964 were the ones held by Bill Fouts, but there were many others. For the Control Programmer team to obtain project direction, there were meetings with R.C. Kellett, the AFRM 009 project engineer. Because the Control Programmer in AFRM 009 was in effect the pilot that was going to command and fly the Command Module, there were requirements for the Control Programmer to work with many other Command Module systems, and to conduct the daily business with the engineering groups of all those systems, there were many internal meetings between engineers from the Automated Systems Design Unit and engineers from other S&ID units. There were weekly management meetings with Autonetics that were held starting early in June. There were internal Space and Information Systems Division Control Programmer design reviews and informal short meetings with Autonetics engineers to discuss problems. There were mandatory meetings with our director, Dave Levine, and periodic Automated System Design group meetings. There were some meetings between Bill Paxton and me; I also had numerous meetings with my team members. There were meetings with supervisors of other groups and phone calls between members of the Control Programmer team and other internal S&ID engineers to discuss methods for implementing interfaces between the Control Programmer and their systems or to resolve any existing problems. There were telephone calls between engineers at the Space and Information Systems Division and engineers at Autonetics to discuss the Control Programmer mechanization or to schedule meetings to solve design problems. There were meetings between members of the Control Programmer team and other engineers in the Automated Systems Design group. There were many telephone conferences calls between the Control Programmer team members and the various AFRM 009 subcontractors. Finally, there were Control Programmer team weekly coordination meetings with the Autonetics team engineers, the first one held in October and then every week thereafter.

Supervisors and lead engineers attended at least two meetings with Bill Fouts each month in 1964. The attendees at those meetings were generally Larry Hogan, Carl Conrad, Bill Paxton, and me. These meetings were where leaders generally got the dates from Bill Fouts for the Control Programmer systems to support the AFRM 009 schedules and to discuss the new AFRM 009 mission timeline and any new changes that NASA had been contemplating. The meetings also served as a sounding board for the topics that were going to be discussed during the weekly management meetings with Autonetics and gave the AFRM 009 team an opportunity to talk about the Control Programmer system problems that had to be solved. Sometimes Bill Fouts would invite people such as R.C. Kellett and Ken Turner, who was one of the NASA resident representatives at the Space and Information Systems Division, to discuss the problems associated with the Ground Control Communications and Control Programmer Radio Command Controller. In those meetings, Ken Turner would, for example, suggest that the AFRM 009 team should communicate more closely with the Ground Control Communications organization. During the Bill Fouts meetings, the team also strategized about methods that could be used to help Autonetics with the problems they were having with procuring parts for the Control Programmer systems.

One of the meetings with Bill Fouts took place after he had attended a meeting in Gary Osbon's office. Bill wanted to let his team know the important topics that had been discussed at that meeting. There were at this time several other Apollo airframes being worked on in Downey with which the Automated Systems Design group was not actively involved. These included AFRM 001 and Boilerplate 14. The Automated Systems Design group had nothing to do with AFRM 001, and for Boilerplate 14 Bill Fouts' department was merely supplying one of the refurbished prototype Control Programmers for installation. It was announced at the Gary Osbon meeting that AFRM 001, AFRM 009 and Boilerplate 14 were Class I vehicles. H.C. Langmore, the senior project engineer, had authorization to provide direction in all areas of Class I vehicles, such as expenditures of overtime and work assignments. Charlie Feltz was appointed as the head honcho on Class I vehicles. This meant that any matter involving a possible change in cost and schedules on Class I vehicles had to be cleared through Charlie. He was a VP who had worked for the company for many years, came to S&ID from the aircraft division and wielded a big stick that demanded respect, was authoritative, and caused much fear.[1]

A few words about boilerplates that I have mentioned above are in order because of their importance to the Apollo program and its success. Boilerplates were models of the Apollo capsule that were flown unmanned; they were of rugged construction and simulated the weight, shape, and center of gravity of the Apollo capsule. They were the workhorses of the Apollo program. There were thirty boilerplates numbered from 1 to 30 in the program under the NASA contract from the Manned Spacecraft Center in Houston. They were used for many ground tests and flight tests, such as being dropped on water with dummies inside to test the capsules' water landing strength and launching some of them by a Convair Little Joe II booster at the White Sands Missile Range in New Mexico to test the operation of the Launch Escape System that was used to propel the capsule and the crew to safety by the launch escape rocket motor attached to the apex of the capsule if anything went wrong with the booster rocket during the launch phase.[2]

While AFRM 009 was in its test phase, with the Control Programmer installed before being delivered to NASA, I was called by Thom Brown to come help the AFRM 009 test team solve a problem with the Control Programmer. Even though I worked all night and eventually located the source of the problem, all the AFRM 009 test team managers resisted my solution, which stemmed from Charlie Feltz's authoritative practices. (I will relate that incident in full in a later chapter.) In addition, Fouts was asked to provide a detailed plan for AFRM 009 with milestones that could be met, a task that was given to me, which I completed by the required date. There was also discussion about seven functions in the Lockheed timer of the Control Programmer that were going to be modified and the problems that were still not resolved with the Communications System and the Control Programmer Radio Command Controller–Ground Control Communications System. Bill Fouts was told at the meeting that NASA had commented on the AFRM 009 Control Programmer mechanization and that they had said, "Autonetics, do it!" In essence, NASA was telling Bill to do anything that had to be done to get Autonetics to deliver the Control Programmer on schedule. Dave Levine had also attended this meeting because it was after the Osbon meeting that Dave called a meeting in his office for what he probably thought would be a planning session on the plans that the Control Programmer team needed to support the Autonetics activities. (This meeting with Dave will be discussed later.)

There were not too many meetings with R.C. Kellett, AFRM 009 project engineer, but these were all necessary due to his position in the AFRM 009 project. These meetings always revolved around the AFRM 009 mission and Control Programmer delivery schedules to make sure the Control Programmer would support the S&ID AFRM 009 schedules. There were other topics that were discussed with R.C. Kellett, but they are too numerous to list here. At one of the meetings, Kellett approved having engineering-to-engineering letters between the Space and Information Systems Division and Autonetics if they were purely technical and did not include Procurement Specification deviations, budgets and funding, or changes in delivery schedules; this served as the limit of my authority in dealing with Autonetics, as will become evident later in this narrative. Most of the business in the AFRM 009 and Control Programmer project was conducted during engineering-to-engineering meetings internal to Space and Information Systems Division. This was the way things got accomplished. I scheduled my Control Programmer team's meetings regularly with members of Larry Hogan's group, which included Carl Conrad's team, and scheduled meetings with engineers from groups of systems that interfaced with the Control Programmer. Meetings were also held with the system purchasing agents and the contracts people in matters regarding the IDWA with Autonetics and the purchase order for the Attitude Reference System.

One of the first meetings I had with Carl Conrad occurred early in the program and involved talking about the extensive redundancy that Autonetics had implemented in the mechanization of the Control Programmer to satisfy the original AFRM 009 mission as reflected in the original Procurement Specification. I had recorded it in my engineering notes because it was my introduction to electronic component redundancy and my real introduction to Carl, who had more experience in the rocket science business than me. I had reviewed the Autonetics design of the Control Programmer in the application of the redundancy design requirements in the Procurement Specification and believed that the redundancy was excessive. As I discussed my findings with Carl and he explained to me the intent of the redundancy requirements in the Procurement Specification, it seemed to me that my assessment of the Autonetics redundancy was not out of line. Carl agreed at this meeting that some of the redundancy had to be reduced and/or deleted. The Control Programmer Procurement Specification had guidelines for the electronic redundancy required in the different types of Control Programmer electronic circuits and was not difficult for me to understand.

In 1964, Carl Conrad was the most knowledgeable engineer on the design of the AFRM 009 Control Programmer. It was Larry Hogan's group's responsibility to provide the Automated Systems Design group with Control Programmer systems requirements, which would then have to be converted to Control Programmer design parameters to be included in the Control Programmer Procurement Specification.

After my meeting with Carl on June 2, 1964, and after we had reached agreement on which redundancy had to be reduced and which should be eliminated, I called Project Engineer Bob Wark at Autonetics and described to him the results of our discussion. I asked Bob to give us a SWAG on the work (SWAG is an acronym for "Scientific Wild Ass Guess"). I told Bob that we needed the SWAG within the next two days. Bob said he thought Autonetics could respond in that time frame and he would call me back the following day and verify what he had told me.

Most of the time when design changes were being considered for the Control Programmer by Bill Fouts' department, I would call Vaughn Slaughter at Autonetics and request that Autonetics give us a SWAG on the design changes that included cost and schedule impact on the system. The SWAGs were important to the Space and Information Systems Division because they gave us a quick look at the impact that the changes could have on cost and schedules. The SWAG had no contractual significance, and Autonetics was not obligated to honor its cost and schedule estimates, but they were good ballpark figures that could be used by us for planning. If the AFRM 009 program could not live with any one of their SWAGs' cost or schedule impact and the design changes were mandatory, a meeting with Charlie Feltz would have to be scheduled to get some relief on cost or schedule, depending on which of the two was out of line.

The other system that the Control Programmer team was responsible for, as mentioned previously, was the Attitude Reference System, and of course meetings on the system had to be attended to arrange for AFRM 009 systems to be delivered to support vehicle testing and vehicle flight. The Attitude Reference System was an off-the-shelf system purchased from American Wianko. This system was to be installed in AFRM 009 and used by the ground controllers to monitor the attitude of AFRM 009 as a backup to the Stabilization and Control System Attitude Reference System. The system was necessary because some of the early abort missions had to be handled by ground controllers via radio command. The Space and Information Systems Division purchasing agent for the Attitude Reference System was Stan Remick. Z. Stanley Kleczko and I met with him many times. Meetings with Stan Remick were also attended by Bob Ellison and Bill Vance from Larry Hogan's group. Equally necessary were meetings with American Wianko to negotiate system delivery schedules and to solve engineering problems specific to the application of the system to the Apollo program, as well as meetings at which American Wianko representatives gave us status reports on their program.

One of the things that had to be determined was how the AFRM 009 Command Module was going to be tested at the Space and Information Systems Division to ensure its readiness for delivery to NASA. This is what we referred to as the "checkout philosophy." To determine this checkout philosophy, Carl Conrad, Steve Parrish and I met with Dick James and Hughes from Ground Support Equipment to discuss and put together a plan. Once we all agreed on how we were going to check out the AFRM 009 vehicle, we left the details to the ground support equipment engineers. It would eventually fall on Thom Brown to support the AFRM 009 Command Module checkout while it was being processed through the factory.

During one of our management meetings with the Autonetics team, they asked us to provide them with information they needed to design their Control Programmer test equipment. One of the automatic functions that the Control Programmer performed just prior to the separation of the Command Module and the Service Module in preparation for Earth reentry was to isolate the Command Module from the Service Module propellant tanks. It accomplished this task by electronically activating a solenoid in the propellant isolation valve. What Autonetics wanted was a complete electrical characterization of the solenoid so they could replicate it in their factory test equipment.

When I returned to Downey, I held a meeting with my team to discuss the Autonetics request and make the team aware of what Autonetics was requesting. Z. Stanley Kleczko told me that the person to see was Nick Glavenich. As I requested, Stanley called Nick and asked him to schedule a meeting with our team and the engineers who were responsible for the propellant isolation valve solenoid. When my team met with Bill Wagner of the Reaction Control System group, we learned that the characterization of the solenoid we were looking for did not exist. Bill told us that they would have to run some tests to get it for us. (The exact mathematical term used for the dynamic characterization is L d-i-d-t and is written L di/dt.) It was agreed at the meeting that the dynamic characterization was going to have a large electrical kickback to the Control Programmer and most likely cause damage to the Control Programmer. Later, due to my team's investigation of the electrical kickback to the Control Programmer, the Control Programmer team realized that there was no protection against this kickback. As the investigation proceeded further, our team found out that there were other electrical solenoids that the Control Programmer had to electronically activate and that there was also no protection against their effects.

The protection that was required in this situation was the addition of diodes on all the solenoids in AFRM 009; unfortunately, despite several meetings between our team and multiple design units that were responsible for the other AFRM 009 solenoids, not one of those units wanted to take the responsibility of adding these diodes because it involved time and money. At various other meetings, my team insisted on the urgent need for the addition of the diodes. It was finally decided that the Electrical Power System department was going to be responsible for adding the diodes to the AFRM 009 vehicle. Our team met with Loy Updegraph of the Electrical Power System department, and he verified that they would take care of the diode additions. The addition of the diodes was mandatory because any electrical kickback from any one of the solenoids could damage the Control Programmer during vehicle testing or during Command Module flight, possibly causing complete failure of the AFRM 009 mission. It is worthwhile to note that the addition of diodes for arc suppression of all solenoids would solve the problem that my team uncovered, but some solenoids (such as those that controlled the reaction control thrusters in the Command Module and the Service Module) could not afford the delay resulting from allowing the full electrical kickback to decay by the addition of diodes across the solenoids. A study concluded that a suppression circuit had to be designed to solve the problem of these special solenoids instead of just adding diodes.[3]

One of the most important interfaces that the Control Programmer had was with the Electrical Power System. The electrical group released wiring diagrams that defined the Control Programmer electrical interfaces. The wiring diagrams were the responsibility of John Chee, with whom our design team met many times to define these interfaces and to obtain wiring diagram copies.

Management meetings with Autonetics were originally requested by Bill Fouts and were held every Monday. The meetings were to be between Autonetics and the Space and Information Systems Division management, but that plan quickly went by the wayside. The only people who knew what was really going on in the design, development, and mechanization of the Control Programmer were the lead designers, lead engineers and lead project people, and even though they were not considered part of management, they had to be included in the management meetings. At the Space and Information Systems Division, Bill Paxton and I would sometimes meet on Friday to plan what was to be discussed with Autonetics the following Monday. Most of the time, the agenda for the Monday meetings included the topics discussed in Bill Fouts' meeting during the week before the Autonetics management meeting. In the first meeting with Autonetics, Bill Fouts' team kept bringing up controversial items, and the most vocal Autonetics person offering his opinion against S&ID items was Vaughn Slaughter. On about Vaughn's third objection, Bill Fouts leaned over toward me and asked, "Who is that guy?" I told Bill Fouts, "His name is Vaughn Slaughter, and he is from the Project Office." Bill's response was "He's good." Vaughn was a strong spokesman for the Autonetics team and soon became one of the persons I talked to regularly.

The first order of business for the Control Programmer team, as mentioned before, was to nail down the Control Programmer mechanization based on the original release of the Control Programmer Procurement Specification so a baseline system could be established. The Control Programmer redundancy that had to be reduced/or deleted had already been discussed with Autonetics. At this first meeting, it was decided that a meeting between S&ID engineers and Autonetics engineers should be held to decide the course of action on the Control Programmer redundancy problem. One problem that Autonetics made Bill Fouts' team aware of during one of our management meetings was that the Lockheed timer was qualified to only 15Gs acceleration, but the AFRM 009 requirement was that it be qualified for 20Gs acceleration. This problem, it was decided, would be given to Chuck Markley of Space and Information Systems Division and Ed McTeague of Autonetics so they could come up with a solution. The problems that Autonetics was facing with building the factory test equipment and possible solutions were also discussed. (This was the meeting at which Autonetics asked us to get the electrical characterization for the propellant isolation valve solenoid that was discussed earlier, for which the addition of protection diodes was required.)

In subsequent management meetings with Autonetics, the cold plate that was needed for the Control Programmer was discussed. The Control Programmer had to be maintained at a specific temperature so it could operate at maximum efficiency and, most important, maintain its high level of reliability. The Control Programmer's temperature was maintained by a cold plate that was placed on top of a platform inside the AFRM 009 capsule; the Control Programmer was then mounted on top of the cold plate. To control the Control Programmer temperature, a mixture of cold glycol and water was circulated through the cold plate. Again, the task of coming up with a solution to the problems with the Control Programmer cold plate was assigned to Chuck Markley and Ed McTeague. Discussed later were the comments that I had made during my review of the original issue of the Control Programmer Procurement Specification regarding the design of the Control Programmer and those comments made by Autonetics that resulted in an agreement on the changes to be made to the baseline Control Programmer. It was at this time that I was given the task of writing IDWA 6513, which covered the new mission design Control Programmer. One big item that Bill Fouts' team stressed to Autonetics in a previous meeting was that the Space and Information Systems Division wanted to reduce the cost of the Control Programmer. Seven cost reduction items were presented by Autonetics in a subsequent meeting, which were taken by the Downey team for consideration.

As was mentioned before, one of the many important tasks that my team had to perform in 1964 was ensuring that the Control Programmer mechanization matched the design provided in the original Procurement Specification. Gene, Don, and I did this by holding informal meetings with John Rowe and his team at Autonetics to discuss the Control Programmer mechanization. The status of the mechanization was also discussed at the weekly technical coordination meetings on Tuesdays. Another method that was used to track the Control Programmer design was reviewing numerous drawings and schematics that Autonetics had put together on the Control Programmer. I communicated with John Rowe, Vaughn Slaughter and Sam Rotuna many times to get these drawings from Autonetics as soon as possible so that my team and I could start reviewing the Autonetics mechanization early and spot any possible disagreement among the drawings, schematics, and the Control Programmer Procurement Specification.

On June 9, 1964, Carl Conrad and I met with Bob Wark, John Rowe, Al Okamura, Chapman, and Bauer of Autonetics to discuss the deletion and reduction of the redundancy in the Control Programmer. At this meeting Autonetics provided Carl and me with three different Control Programmer redundancy mechanization candidates. Each of the three options had a different probability of success. These candidates would have to be presented to the Reliability group back in Downey so they could help us choose the best option that would meet the required Control Programmer probability of success for AFRM 009. Another meeting with Carl and myself included Harry Lewis, Chuck Markley, W.F. Dyer, and M.G. Carmichael to review the original release issue of the Control Programmer Procurement Specification. At this meeting, we decided that the changes that had to be made to the original issue of the Procurement Specification because of the specification review a few weeks earlier, plus the changes that had to be made due to the AFRM 009 original mission change to the new mission, together formed the "A" revision issue of the Procurement Specification. Chuck Markley in the Automated Systems

Design group was responsible (with my help) for making all the agreed-to changes to the document, and it was Chuck's job to ensure that the Procurement Specification, with all the appropriate changes, was formally released. In such a case where the Control Programmer Procurement Specification was being modified from a "no change" revision to an "A" change revision, Autonetics was given some time to review the alterations to the Control Programmer design and then come back to S&ID with comments on design changes that they could not respond to or those changes that they felt would impose a great strain on them. These differences would be negotiated to reach a reasonable compromise, at which time Autonetics would approve the "A" revision Procurement Specification with minor revisions. Approval of the Procurement Specification by Autonetics was then followed by budgeting and pricing. The budget and pricing was then presented to Bill Fouts' department at one of the Autonetics management meetings. The final formal submittal to the Space and Information Systems Division was a firm cost proposal that the Downey financial experts would later negotiate with their Autonetics counterparts.

There was an IDWA group at the Space and Information Systems Division, and I met with Dick Cordon and Tom Whaling of that group regularly at the beginning of the program. The responsibility of writing the IDWA to replace the minor version that Autonetics had been working under was given to me; the number was IDWA 6513, and it spelled out the contract between Autonetics and S&ID. Putting this document together took many meetings with different people, along with a lot of coordination and document reviews. This process was intense and took most of my time because it had to be released to Autonetics as quickly as possible. At the same time, I had to give assignments to the engineers on my team and help them out whenever they needed assistance. Autonetics was crying for the IDWA, and upper management kept pushing me to get it done fast. Once IDWA 6513 had been written in a rough draft, I met with Tom Whaling to categorize all the topics and put each category in the proper paragraph. The rough draft of IDWA 6513 was then submitted for typing. When it came out of typing, I reviewed it for accuracy, errors redlined on the hardcopy with corrections; when I was done, I submitted the proofread and corrected copy for final typing and requested that two or three carbon copies be made on the final typing. Once the final copy and the carbon copies were ready, I circulated the carbon copies for review and comment. I personally took the review copies to Bill Fouts, Bill Paxton, Larry Hogan, Carl Conrad, Tom Whaling, and R.C. Kellett so that review would be expedited. Everyone understood the urgency, so cooperation was excellent, and comments came back quickly. All the comments were rolled into the IDWA and quickly submitted to be retyped. I obtained engineering signatures on the retyped copy and personally delivered that copy to Tom Whaling. From there it was out of my hands. Tom would send IDWA 6513 formally to Autonetics for review and acceptance. After Autonetics completed its review of the IDWA, there followed a negotiation period. By June 16, 1964, IDWA 6513 was already released.

5

AFRM 009 Control Programmer Supervision

An important feature that had to be incorporated into the Control Programmer and into AFRM 009 was the capability of testing the Control Programmer while installed in AFRM 009. The responsibility of ensuring that this happened, as mentioned previously, belonged to Thom Brown. Thom would coordinate all test requirements with my team, with the Autonetics engineers, and with Paul Rupert of the Building 290 AFRM 009 test team. Because of his responsibility, Thom was deeply involved with the Control Programmer design activity, joining in all our design reviews, and he was always aware of the status of the Control Programmer design. One day, early in June 1964, when Thom was going to Autonetics to meet with one of his counterparts, I asked him to request copies of various Control Programmer drawings from Sam Rotuna or John Rowe. When Thom returned to Downey from Autonetics, he told me that Autonetics had denied his request, claiming that IDWA 6513 made no provisions for them to supply us with those drawings. I told Thom that they were full of crap because I had specifically put an item in the IDWA stating that they must provide S&ID with copies of the drawings I had requested. I even quoted Thom a paragraph number from the IDWA. I put this item on the agenda for the next Monday management meeting. Many times, subcontractors refused to give the prime contractor drawings of their products, claiming that they contained proprietary information. Autonetics finally relented and gave us copies of all the drawings that I was requesting. These schematic drawings were then used by my team to verify the Control Programmer design.

Pulling from what I had learned from my Hound Dog cruise missile mentor Ed Kelley, I gathered my team in a conference room, took the Control Programmer drawings we had received from Autonetics (as well as a copy of the Procurement Specification and some colored highlighters) and sought to verify the Autonetics design. Our team started with the paragraphs on Control Programmer design requirements in the Procurement Specification; we colored the first paragraph using a specific marker and then found the part of the Control Programmer System where the same paragraph design had been mechanized, coloring it with the same marker. My team continued in this way with every design requirement paragraph of the Control Programmer Procurement Specification. If a design requirement in any paragraph had not been mechanized or was improperly mechanized, we would not color that requirement. When we were done, we could tell whether the mechanization had been done correctly by tallying the number of design requirements in the Control Programmer Procurement Specification that had not been colored. A correct Control Programmer mechanization was indicated by a

Procurement Specification that had all design requirement paragraphs marked. A meeting with Autonetics followed to discuss the findings of this review.

On June 19, 1964, John Chee from the electrical group released the AFRM 009 schematic diagram, V14-945001, which depicted the Automatic Command Controller and the Radio Command Controller of the Control Programmer with all the wire connections to each of the Control Programmer's interfacing systems. This drawing was a long-awaited milestone for AFRM 009 because it was the baseline for the AFRM 009 Control Programmer interfaces and could be reviewed by those whose system was involved in the Control Programmer's operation. Because I was acting supervisor for Bill Paxton at the time, I signed for the approval to release the drawing.

The relationship that I expected to have with Bill Paxton was one like what I'd had with Ed Kelley on the Hound Dog cruise missile program, but it never really materialized. Ed was always interested in the technical end of the job as well as the administrative end, and he was a great leader. He wanted to be in the middle of a solution to any problem, and he was always ready to discuss any challenge, regardless of what it might be. Ed constantly provided words of encouragement or welcome hints on solutions. Sometimes he would say, "Let's go see the chief engineer or the project engineer to get an okay on a problem solution." Or Ed would say, "Why don't you go talk to such or such a person to see what he thinks?" There was always some kind of supervision given by him, no matter how slight, and we knew that any decision we reached on our own would be supported by Ed. During our days on the Hound Dog cruise missile program, Ed spent little time in his office during working hours; instead, he was always helping us work a problem or wanting to know how our solution was coming along. To catch up with his mail and other administrative duties, he would usually stay late at work each day and come into work on Saturdays or, on rare occasions, sometimes even Sundays.

Bill Paxton was the polar opposite of Ed Kelley. He spent all his time in his office, oblivious to what was going on outside, and most of his time was spent rewriting internal letters that members of my team or other members of the group had written. It got so bad that one day Don Farmer was going to refuse to approve a letter that he had written after Bill completely rewrote it. Don said to me that the rewritten version was not the letter that he had originally written. The whole business of having my team's letters completely rewritten bothered me a hell of a lot because it was not contributing to the program one bit and only serving to put a bureaucratic delay in what my team was working to accomplish, as well as undermining team morale. However, I thought if that made Bill Paxton happy and my rewritten letters said the same thing that I wanted to say, I was not going to waste my valuable time arguing over it. I figured that the English language is such that there are a thousand ways to say the same thing, and if Bill wanted each letter to say it the way he wanted it said, then so be it. I had real work to do and was not going to let a rewritten letter derail me. My opinion was that Bill was our supervisor and he just was not supervising us and was useless as far as contributing to the program was concerned. Because of Bill's obsession with rewriting letters, many of us started to say he was like an English teacher, and eventually everyone referred to him as "Teacher" behind his back.

Not having known Bill Paxton previously, I had no idea of how he operated as a supervisor or what real engineering or management capabilities he possessed, so I began by walking into Bill's office and briefing him on what was happening on the program, thinking that he would be interested, but after I was done with my report, Bill would give me suggestions that sounded like he didn't understand what I had said, as they were counter to the solution I was trying to find or did not reflect sound engineering judgment. Beyond that, Bill often changed the subject and would always give me some action items to work on that had little to do with what my team was doing. Completing the action items that I had been given took valuable time away from my team that could have been used more effectively on the program. Eventually, I stopped going into Bill's office on a regular basis and went to talk to him only when I thought it was necessary (and sometimes not even then).

This is a good time to discuss the Control Programmer Qualification Test. Every system that went into an Apollo Command Module had to pass a series of environmental qualification tests. Because the Apollo Command Module was going to go into space and would be subjected to high vibrations, high accelerations, high temperatures, low temperatures, high shock, and high humidity, it was mandatory that all systems be designed to work properly under these conditions. All these environments were spelled out in the Procurement Specifications for each of the different Apollo systems. The subcontractor of each system would first be required to submit a plan to the Space and Information Systems Division for approval on how they intended to run all the environmental tests on one of their systems that was of production quality (that is, the same as the one that was to fly in the Apollo capsule). When the subcontractor's plan was approved, the subcontractor would then write a Qualification Test Procedure, which also had to be approved by S&ID. On my team, Gene Rucker and Don Farmer were coordinating with Autonetics on writing the plan and test procedure and reviewing and approving the final version of both documents. The Qualification Tests on the Control Programmer were going to take place in 1965 in the Autonetics laboratories. My envisioned plan was that our three team members would support the tests on a twenty-four-hour basis on a three-shift rotation. This approach was necessary because of how critical the Control Programmer was to the success of the Apollo program, but was it something that management would approve? The only way to find out was to test the waters by presenting the plan to management, starting with Bill Paxton and then Bill Fouts.

In our meetings with the Autonetics managers and engineers, we had approved their schedule for delivering Control Programmer systems to us. Autonetics also provided our team with a schedule for the qualification testing of the Production Control Programmer system identical to the one that was going to be delivered to us for AFRM 009. I knew that this testing period was critical to the program because this was when the subcontractors would uncover problems, and sometimes they would need verbal agreement or a written approval of the method of solution for any of their problems. As mentioned earlier, my team paid attention to every detail in the program and left nothing to chance. So, with the AFRM 009 schedule being so tight, and given the incentive dollars that the company stood to earn for an early delivery, I figured that the program could not afford to lose valuable time. Besides the incentives, there was the matter of timing in terms of getting to the moon before the end of the decade and the national prestige at stake that hinged on timely execution of all actions that led to the completion of Apollo milestones. This testing at the subcontractor was generally done around the clock at the subcontractor's facility, as would be the case for the Control Programmer. I felt strongly that my team had to support this testing at Autonetics during their twenty-four-hour activity.

Knowing all this about the Control Programmer testing, I put together a schedule of how the team was going to support the testing effort. I assigned time slots to Gene Rucker, Don Farmer and myself for twenty-four-hour support during the Control Programmer testing period. The schedule for the plan that I had written on a large sheet of paper had the names of all the Autonetics managers and engineers with their work and home phone numbers. It also had my team's home phone numbers and the phone numbers where we could be reached during any twenty-four-hour period. In addition, my plan required Gene and Don to call me at home any time of the night when they were on duty to let me know whether I needed to come down to Autonetics to help them solve a Control Programmer problem or to provide Autonetics with written engineering direction. I made several copies of the schedule, thinking of making enough copies available so that I could give one to any S&ID member of management who was involved with the AFRM 009 Control Programmer. Because my plan involved using numerous group personnel, I went into Bill Paxton's office to let him know the details of what I had in mind. When I presented the plan to Bill, he rebuked it and said it was not needed. So, I figured that if he thought the plan was unnecessary and did not want to go present it to Bill Fouts, the hell with it. I returned to my desk and filed the plan in my desk drawer.

About three or four weeks later, just after Bill Fouts had attended the meeting with Chief Engineer Gary Osbon on AFRM 009 that was mentioned earlier, Bill Paxton came out of his office and headed straight for my desk. Bill was shaking like a leaf, and his hands were trembling. He told me that our director, Dave Levine, wanted to see Bill Fouts, him, and me in his office immediately because he wanted to know how we were going to support Autonetics during the AFRM 009 Control Programmer Qualification Testing period. I opened my desk drawer and pulled out the copies of the plan and schedule that, according to Paxton, had been unnecessary. I then followed Bill Fouts and Bill Paxton to Dave Levine's office, which was located one floor above Fouts' office. It took me several minutes to catch up with Bill Fouts and Bill Paxton (even though Bill Fouts was of medium height and slender, he walked exceedingly fast and took the stairs two at a time—it was amazing that a small guy like him could walk so quickly). I caught up with the two of them just as they were entering Dave Levine's office. Now, Dave had a reputation for abrasively wire brushing those who met with him in his office and who were ill prepared to answer his questions. In his meetings, Dave also looked directly at the person he was speaking to, and if you were not the one Dave was addressing, you did not dare butt in because he would not look at you to acknowledge your verbal input. You spoke only when Dave spoke to you. It was a very traumatic experience to meet with Dave when you were not prepared to discuss whatever subject was on the agenda. Dave was also famous for holding meetings during the lunch break, which was referred to by all as "having lunch with Levine." Stories about these two types of meetings circulated through the company grapevine, and none of them ended on a good note, which was why people feared to meet with him (and which explained why Bill Paxton was shaking).

Dave was in his office, waiting at his desk, when the three of us walked in. As soon as we were seated, the first thing that Dave did was look at Bill Fouts and ask, "How do we plan to support Autonetics during the Control Programmer testing period?" Bill Fouts turned around and, looking at Bill Paxton, said, "Bill?" as if expecting an answer from him. I had turned my head around and was looking at both Bill Fouts and Bill Paxton. Paxton was still nervous and shaking a bit. When Dave turned to look at Paxton, Paxton turned around and looked at me and said, "Jim?" I slowly turned around to look at Dave, and when Dave turned and looked at me directly, I very calmly pulled out three copies of the plan and schedule and handed one to each of them. Looking directly at Dave, I explained to him in detail all the things that were in my plan and how the support was going to be handled according to the schedule. I told Dave that during the shifts when Gene and Don would be supporting the testing at Autonetics, they had instructions to call me at home at any time. They could call me for consultation, and if I were needed at the Autonetics facility to provide direction or sign a document, I would drive down there immediately. The four of us discussed the plan for a while, and Dave said that it had his approval. Fouts and Paxton agreed with Dave that the plan was good. The three of us stood up and started to walk out of Dave's office; that was when Dave looked at me straight in the eyes and said very distinctly, "Remember, De La Rosa, *you* own Autonetics." I took those words from Dave as marching orders.

From that day forward, I very seldom went into Bill Paxton's office unless it was to attend a meeting to which I had been invited or to participate in a conference telephone call on which Bill needed my presence. I solved most Control Programmer System problems using my engineering experience and judgment. At times when I thought that Bill Fouts should be made aware of something important, I would go talk to him directly in his office without even consulting with Bill Paxton. Months later, after thinking and agonizing over the fact that I was making all decisions on the Control Programmer without Bill Paxton's help and that the supervisor was sitting in his office every day rewriting letters, oblivious to what was happening, I thought I would devise a way to get Bill involved. My plan was to go into Bill's office and let him know every time that a decision on the Control Programmer had to be made and ask him to give me direction. When I implemented the plan, it backfired, and I had to give it up because Bill would procrastinate for days and do nothing. Ironically, when decisions were not forthcoming in a timely manner and things went awry, I was blamed for it. I had no choice but to go back to my old method of operating because Bill Paxton had *no* idea what was going on in the AFRM 009 Control Programmer program.

So why did I take all this crap? Because I was not a quitter. I believe that when given responsibility and a level of authority, you should exercise your authority to fulfill your responsibility by making decisions even if they are risky; if you cannot do that, you do not belong there, so get out.

Bill Paxton's inability to supervise was so evident that Carl Conrad disapproved of him, and Carl always let us know how he felt through the negative comments he made about Bill. Carl would say that Bill Paxton was worthless and that Bill Fouts should get rid of him. Carl, however, never expressed his feelings directly to Paxton.

While I was at Autonetics one time early in the program in 1964, one of the Autonetics engineers told me that Autonetics needed some type of engineering direction. I asked Bob Fettes, the engineering supervisor, what they required, and he told me verbally in detail the engineering direction that Autonetics needed from the Space and Information Systems Division. To bypass the Bill Paxton delay, I wrote a letter in long hand with all the details that Bob had given me and stated in the letter that S&ID was authorizing the engineering direction. Bob's secretary typed the letter and I signed it, and Autonetics moved ahead with the engineering direction. The letter had served its purpose, which was to let Autonetics continue its work immediately without interruption. Autonetics management was smart and accepted my letter as S&ID authorization because Bob Fettes knew that if the letter had to come from Bill Paxton or Bill

Fouts, it would take weeks to get out of Downey. When I wrote that first letter of direction to Autonetics, I was acting in accordance with Dave Levine's instructions that I owned Autonetics. After I returned to Downey with copies of the letter, I gave a copy to Bill Fouts and Bill Paxton; neither of them challenged my authority to write letters of engineering direction to Autonetics on the spot. From that day forward, this became my modus operandi with Autonetics.

In the ensuing days after I returned with that first letter of direction that I had written and signed at Autonetics, Bill Paxton naturally rewrote the letter in its entirety. A week or so later, I received the rewritten letter from Bill, which I was expected to approve by initialing it at the bottom next to my name; Bill's name was at the bottom of the letter so he could sign it, giving the illusion that the direction came from him. I held on to the letter for several days, not wanting to approve it because in its rewritten form it was not my letter anymore. I even considered confronting Bill to tell him why I was not going to initial the letter. However, the more I thought about it, the less important it became, so I eventually gave my approval to the rewritten letter. Of course, by that time the rewritten letter was academic and served no purpose. The original letter had already permitted the necessary action to take place at Autonetics, and arguing about it would be a waste of valuable time; doing so would also tend to anger and aggravate me, raise my blood pressure and be detrimental to my health, and for what—a worthless piece of paper? Thereafter, every engineering letter of direction written by me and given to Autonetics on the spot was rewritten by Bill Paxton, and several days or a week later I would initial the rewritten copy without even reading it.

After the incident with the letter of direction to Autonetics, I started to take more authority in the things I did. Someone asked me one day, "Jim, how do you know if you have or do not have the authority to give direction or to authorize things to be done?" The answer I gave was "I just assume that I have the authority, and if I don't, someone will tell me." Well, it happened one day. Larry Hogan called me into his office demanding to know why I had authorized a certain engineering action to take place. I told Larry that I had authorized it because I needed it done to keep from interrupting the work my team was doing. Larry told me it was his responsibility, and he would take care of it. I told Larry it was okay with me, providing I was guaranteed that my team's work was not stopped. There were no hard feelings between Larry and me because of this incident. As the saying goes, "It's better to ask for forgiveness than to ask for permission."

In the beginning, whenever Bill Paxton was gone from the facility, he would leave me as acting supervisor for the period that he was absent. This happened on May 22, June 19 through June 28, and September 4, 1964. As time went on and I got deeply involved in the technical end of the Control Programmer, and more involved with the engineers and managers at Autonetics and S&ID, it seemed that Bill and I drifted further and further apart and had no common ground for discussion. Eventually in 1965, Bill would alternate the acting supervisor assignment between other members of the design unit and me.

The very first and only group meeting held by Bill Paxton was a prelude of things to come. The topics discussed in that meeting concerned the filing system that was going to be set up and how Bill wanted it to be handled. There was nothing about the status of the design of the AFRM 009 Control Programmer or how Bill was going to run the group or what was expected of us: zero supervision. Also discussed were the letters that were being written by members of the group for his signature; regarding these, Bill specified that instead of each engineer signing the final copy of the letter, each letter writer was to initial the letter only, while he would sign at the bottom of the letter. (The letters of engineering direction that I was writing at Autonetics on the spot were never discussed; I continued signing these without Bill's approval signature.) The meeting went on in this way, and we never talked about the problems anyone was encountering or whether anybody needed some help in any of the work we were doing. Bill Paxton never filtered anything to the group that came down from our meetings with Bill Fouts or our management meetings with Autonetics. I was the one who always got my team together and let them know what was happening after every meeting I attended with Bill Fouts or with Autonetics management.

It was on July 17, 1964, that Deke Slayton (one of the original seven NASA astronauts and now the assistant director for flight crew operations at the Mission Space Center in Houston) announced the assignments that had been given to the various astronauts. Everyone at S&ID was excited, even those of us who were working on the unmanned Apollo capsule, because now we knew the astronauts with whom we would be working on the Apollo program. Alan Shepard had been named chief of the Astronaut Office, while L. Gordon Cooper was to head the Apollo Group, with his crew being James A. McDivitt, Charles Conrad, Frank Borman, and Ed White. The astronaut selected to head the Gemini Group was Virgil (Gus) Grissom, and his astronauts were John W. Young, Walter M. Schirra and Thomas P. Stafford. Neil Armstrong was to head the Operations and Training Group, its members being Elliot M. See, James A. Lovell, and M. Scott Carpenter. The week before July 17, S&ID hosted the NASA review of the Apollo Mockup. In attendance for the review were James McDivitt, Mike Collins, Gordon Cooper, and Alan Shepard.[1] It was always evident when the astronauts were in town because we would invariably meet them walking through the halls of the main buildings at work in Downey.

6

Finalizing AFRM 009's Control Programmer on Paper Design

Some important things happened in 1964. July 29, 1964, marked the date that Bill Paxton released the new organization chart as an attachment to an internal letter titled "Unit Organization—Automated Systems Design." The chart assigned the following responsibilities: The first block of the organization chart showed me as lead engineer for the ACS 009 Design, with members of my team being Z. Stanley Kleczko (responsible for ARS Development), Gene Rucker (responsible for Control Programmer Test and Bench Maintenance Equipment Development), and Don Farmer (responsible for Control Programmer Design). Pat S. Erlich was lead engineer for Design Support, with team members Thom L. Brown (responsible for Ground Support Equipment Interfaces), Chuck W. Markley (responsible for Mechanical Interfaces), Milt W. Swan (responsible for Documentation, Schedules, Configuration Control, and so on), and Steve W. Parrish (responsible for Process Specifications, Simulation Support, and S-V Dart Preliminary Design). (The S-V Dart, short for Saturn Five Dart, was the start of the design of the Mission Control Programmer for the remaining unmanned Apollo capsules, AFRMs 011, 017 and 020.)

Pat Erlich was someone whom Bill Paxton hired out of Autonetics (at least this is what Bill told me). Bill told me that Pat was not happy with his position and level of authority. Pat wanted to command more people, but Bill would not yield to him. Pat did not last long—he was gone as quickly as he had come. The reason for his departure was never disclosed, but this type of thing happened often on the Apollo program. Some people in the program were empire builders, doing little of the important work, their only ambition being to have a big block in the organization chart with a lot of employees reporting to them. Then there were the very few managers and engineers who carried the brunt of the responsibility and did most of the work. Pat Erlich had the largest block in Bill Paxton's organization chart and gave the impression that he was some type of empire builder.

The third block in the new organization chart showed Mitch F. Wieser as lead engineer for the S-IB Dart Design. Mitch had John Kelleher (responsible for Development), W.T. Wong (responsible for Test and BME Development), and Frank Chee, John Chee's brother, (responsible for Design). Mitch had the Mission Control Programmer for AFRMs 011, 017, and 020.[1]

I was not involved with the Mission Control Programmer for AFRMs 011, 017 and 020 at this time. There were initially two separate Mission Control Programmers—one named the S-V Dart, intended to interface with the Saturn V booster with AFRMs 017 and 020 as payloads, assigned to Pat Erlich, and a second one, the S-IB Dart Design, that interfaced with Saturn IB booster with AFRM 011 as payload, which was assigned to Mitch F. Wieser. Eventually the two programmers were combined to form one Mission Control Programmer under the responsibility of Mitch.

To handle the AFRMs 011, 017 and 020 Mission Control Programmer (everyone called it the MCP) work, Bill Paxton had hired Mitch Wieser as lead engineer. Mitch had his desk and conference table next to me, and because of that, we talked with each other quite a bit. Mitch was attending classes at some university toward a PhD. I sensed that Mitch had reacted differently than I had to working for Bill Paxton and with Autonetics. We never talked seriously about our jobs and responsibilities, but I sensed that Mitch was not too happy with his position. Instead of saying good things about the MCP, Mitch would make unkind and somewhat caustic remarks against the program. One of Mitch's favorite remarks to make to me whenever some milestone in the MCP project was reached or a negotiation was ongoing with Autonetics was "Sell the sizzle, sell the sizzle"!

Eventually, during the second or third management meeting with Autonetics, the discussions moved to the new AFRM 009 mission. My team and I had been busy revising the Control Programmer Procurement Specification to the "A" letter revision, which included the new AFRM 009 mission and some other changes that had to be made to the Control Programmer for the new mission to work properly. At this time, I gave Autonetics a Control Programmer block diagram that depicted the new mission and that would be included in the "A" letter revision of the Control Programmer Procurement Specification. The block diagram identified the changes to be made to the baseline Control Programmer to bring it in line with the new mission. I also provided Autonetics with the

Enclosure (1)

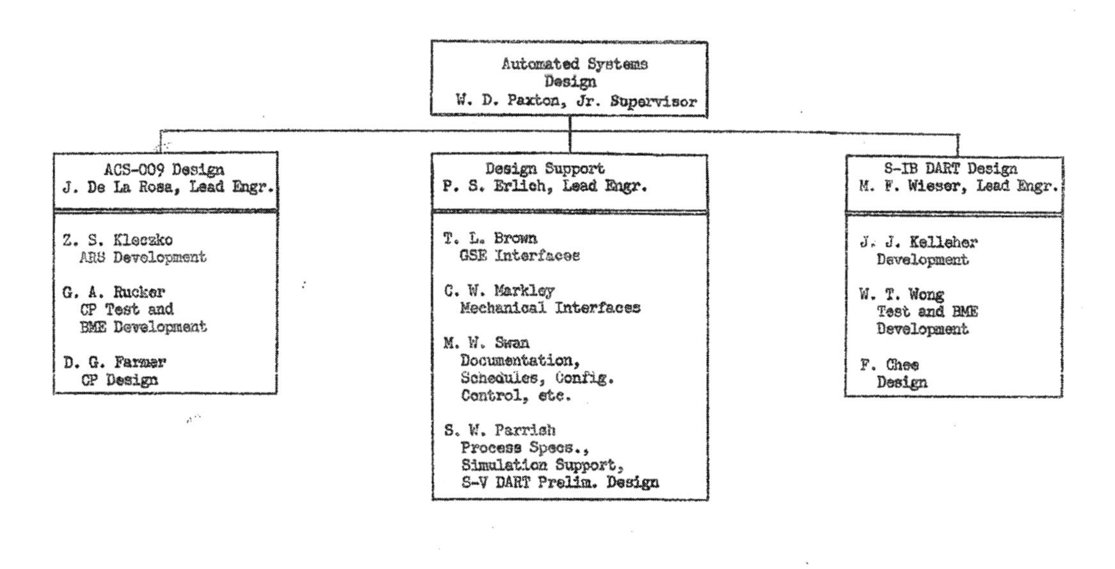

AUTOMATED SYSTEMS DESIGN
UNIT ORGANIZATION

Design unit organization chart for July 29, 1964 (Courtesy of the Boeing Company).

new mission timeline that identified and described each event and listed the time that the Control Programmer had to execute each event during the new mission.

By the beginning of October 1964, my team had all the Control Programmer drawings and schematics for the new AFRM 009 mission based on the "A" letter revision of the Control Programmer Procurement Specification. John Rowe had given me a preliminary copy of the AFRM 009 Control Programmer Mechanization Schematic Rev A that was eventually released on October 28 with the design agreed to by Autonetics and my team, with all alterations made because of my team's internal design review. The drawings and schematics were delivered to us formally by Autonetics by way of the IDWA group, which was a big milestone for the AFRM 009 Control Programmer project. To come to the design agreement, I had scheduled an internal design review on the Control Programmer Rev A configuration on October 6, 1964, in Conference Room 101. The attendees at the design review were Gene Rucker, Don Farmer, Z. Stanley Kleczko, Thom Brown, and me. On the day of the review, I brought the Control Programmer Procurement Specification, all the drawings and schematics of the Control Programmer, and an assortment of colored markers to the conference room. The procedure for verifying the Control Programmer design to the Revision A Specification was the same one that had been taught to me by Ed Kelley and that had been used previously to review the baseline Control Programmer design.

The work on Apollo became very intense at this time and continued all throughout the program. I was so involved in the work, sometimes being on call 24 hours a day, that it was impossible to separate my private life from my work. During this time, NASA had been meeting with Bill Fouts and other members of the Space and Information Systems Division management to discuss further changes NASA wanted to make to the AFRM 009 Control Programmer. Because of all the NASA-proposed changes and the activity at Autonetics, Bob Fettes (Autonetics Control Programmer supervisor) and I decided on October 2, 1964, that S&ID and Autonetics should have engineering-to-engineering technical coordination meetings once a week. It was agreed that the meetings should be held on Tuesdays and be alternated each week between Anaheim and Downey.

The company announced that October 3 was going to be Family Day at our division. On that day, the company was open for visits from employees and members of their families. There were big preparations made for that day, with displays set up for family members to see, and many of the areas were open to all (except those areas where company proprietary work was going on or where it was dangerous for children). I took my family with me to enjoy Family Day at work, and a good time was had by all.

In the management meeting with Autonetics on October 5, 1964, we finally presented to Autonetics the Control Programmer changes authorized by NASA that would make up the "B" revision letter of the Procurement Specification. There were at least twenty changes to be included in the "B" revision. In one of the earlier management meetings, Bill Fouts' team (at my insistence) had requested that Autonetics issue badges to my team that would allow Gene Rucker, Don Farmer, and me to enter the Autonetics buildings where AFRM 009 Control Programmer work was taking place. In September, Autonetics issued the three members of my team picture badges that allowed us to enter the buildings where Control Programmer work was happening without having to go through the lobby to obtain an escorted visitor's badge and then wait for someone to pick us up and take us wherever we went to conduct our business.

Also during the October 5, 1964, meeting, Autonetics presented a Control

Programmer delivery schedule that was contingent on approval of the "B" letter revision of the Procurement Specification. The prototype Control Programmer would be delivered on March 1, 1965, and a Production Control Programmer on April 9, 1965. The rest of the management meetings until the end of the year were spent discussing the changes to the "B" letter revision to the Procurement Spec, as well as the Qualification Test and the environments to which Autonetics was required to subject the Control Programmer in its laboratories.

My very first merit pay increase while on the Apollo program came on October 11, 1964. A Rate Change Notice was presented to me on that date that stated, "Mr. De La Rosa is lead engineer for a sub-unit responsible for development of the Automated Control System for AFRM 009. His enthusiasm towards his assignment is outstanding and his performance is consistently excellent. He displays a remarkable degree of initiative and has developed commendable traits of leadership and organizing capability. Mr. De La Rosa's outstanding level of performance and productivity merit this increase." The Rate Change Notice was signed by W.B. Fouts, Chief, and D.S. Levine, Manager.

Many of the department managers and chiefs used staff people to help them in their administrative duties. During the 1964 period, Bill Fouts had Joe Ferrari, then Bill Fordiani and then Milt Swan to help him. Joe Ferrari and Bill Fordiani were not on Fouts' staff for too long, but Milt Swan was there for the long haul, and he and I became good friends because, organizationally, Milt was in Bill Paxton's group. Milt was a retired Navy commander and was always ready to help. My first recollection of Milt was on November 9, 1964, when Bill Fouts asked him to help my team by writing an Interdivisional Engineering Change Proposal to be transmitted to Autonetics that described some changes we wished to make to the Control Programmer. From that day forward, Milt would write many of the engineering documents that were necessary for the support of the Control Programmer, and he and I would meet so that I could provide the words that had to be entered into the document. Milt was good, which explained why Bill Fouts held on to him for so long. On one occasion, a group of AFRM 009 engineers was meeting with Bill Fouts in one of the conference rooms; some of the attendees were George Cortes, Jack Jenson, Paul Garcia, Mert Stiles, Carl Conrad, Larry Hogan and me. I was sitting on the side of the conference table facing the open door, and as I looked up, I saw Milt Swan pass by; as he paused at the open door, Milt looked my way, and we acknowledged each other with a head nod. After the meeting, I looked for Milt and asked him, "Were you looking for me, Milt?" His answer was "No, I was just looking into the room, and I was just thinking to myself, *Wow, there is a lot of moxie in that room*." (*Moxie* is a word that was used extensively in the old days to indicate someone's great skill and know-how.)

On November 11, 1964, I reviewed the vellum of the final version of the Control Programmer Procurement Specification Revision "B" and approved it for Bill Paxton's signature. I also wrote a rough draft of the Engineering Order to formally release the Control Programmer Procurement Specification and submitted it for typing on November 12. A few days later, the Control Programmer Procurement Specification was released into the company document system. The reason why I was responsible for the final review of the Control Programmer Procurement Specification before its release was that I was the one who had made all the changes that went into the Revision "B" Specification. I had completed this task over a long period of time by taking a copy of the Revision "A" Specification and redlining all the changes that had been authorized by NASA and all those changes that had to be made so that Autonetics could mechanize the Control Programmer in good order. The typing and the verification to ensure that all the redlines were incorporated into the "B" Revision of the specification were Chuck Markley's responsibility. After I had reviewed the final vellum for any letter revision of the Control Programmer Procurement Specification and approved it, it was ready for release. Bill Paxton's signature to the Procurement Specification and the releasing Engineering Order were merely a rubber stamp.

By November 12, 1964, Bill Fouts' team had begun to channel a lot of the communications with Autonetics through Don McLean. Don was now program manager at Autonetics because, with the addition of the AFRMs 011, 017 and 020 Mission Control Programmer, they were expanding their operation. S&ID had given Autonetics an IDWA and a Mission Control Programmer Procurement Specification for the additional work. At the same time, Bill Paxton was staffing up with engineers to work on the Mission Control Programmer (as evident in the new organization chart), the equivalent of the Control Programmer, which would automate the mission for AFRMs 011, 017 and 020. The MCP IDWA was written by someone in Bill Paxton's group, and all the requirements for the three spacecraft had come from Larry Hogan and Carl Conrad. Mission planning and timelines for the new spacecraft had come from Tom Shanahan in the Mission Design Group. All the mission information and the requirements coming to Bill Paxton's design unit from Larry Hogan's group had been put together by Chuck Markley in Bill's group to write the MCP Procurement Specification. Similar to the Control Programmer, the Mission Control Programmer was originally called Mechanical Man, but its name had been changed at NASA's request. The missions for spacecraft 011, 017 and 020 were much more complex and diverse than AFRM 009 and required a more complex automated system, such as the Mission Control Programmer (which I will address later).

Close to the end of 1964, between October 16 and November 16, Z. Stanley Kleczko was gone from the program. Bill Paxton had a candidate to replace Stanley. Bill called me into his office and asked whether I knew Phil Woltman. I told him I knew Phil because I had worked with Phil on the Hound Dog cruise missile program. Bill asked me what I thought of Phil as a replacement for Stanley, and I replied that he seemed okay. Back on the Hound Dog cruise missile program, Phil was a good, reliable worker. Bill hired Phil and assigned him to my Control Programmer team to replace Stanley in working on the Attitude Reference System.

Our Christmas present from Autonetics in December 1964 came in the form of Revision "B" to John Rowe's AFRM 009 Control Programmer mechanization schematic, which contained all the agreed-to changes in the newest revision of our Control Programmer Procurement Specification and had a release date of December 10. This was a big milestone for the AFRM 009 project, and our team immediately informed the NASA subsystem project manager, Gene Holloway, of the mechanization schematic release. Now all we needed was for Autonetics to deliver a Control Programmer System to us so that my team could start the testing that was on the schedule for AFRM 009. The question was: When was this really going to happen?

7

Control Programmer Initial Tests and Eagle Knight Cruz Mora

The next two years, 1965 and 1966, were crucial ones for the Apollo program and more so for AFRM 009 because it was the first flyable Apollo capsule to take flight. In 1965, Autonetics delivered the Control Programmer hardware to us, which was followed by preparation for delivery of AFRM 009 to NASA. In early 1965, there were also changes in management within the company. Sandy Falbaum, who was previously program manager for the Hound Dog cruise missile program in Tulsa, Oklahoma, was transferred to the Apollo program as assistant program manager to Dale Myers. Russ Belles was named as Sandy's replacement as Hound Dog cruise missile program manager.[1] The ongoing Paraglider program also made some management changes: Bert Witte, assistant program manager, was elevated to acting program manager. George Jeffs, former S&ID vice president and Paraglider program manager, was appointed executive director of engineering in the Corporate R&E Organization.[2]

On January 19, 1965, my team on the Control Programmer experienced a panic that could have turned out to be a big setback for AFRM 009. There was a stop order issued at American Wianko on the Attitude Reference System because of quality assurance problems. We were alerted by American Wianko of the stop order through a message sent to Stan Remick, the S&ID purchasing agent for this system. Stan relayed the message to Phil Woltman, who passed it on to me. I went to see Bill Fouts to make him aware of the gravity of the problem, and he quickly formed a team to address the situation. This meant that the Attitude Reference System production line had been virtually stopped until the problems were resolved. As mentioned before, delays of AFRM 009 could not be tolerated because of schedule constraints imposed on its delivery by Charlie Feltz and the incentive dollars tied to the delivery schedule. This was a problem of major proportions that had to be resolved immediately because, if allowed to adversely impact the AFRM 009 schedule, Charlie Feltz was going to come down on Fouts' department like a hammer on a nail.

That same day, early on January 19, I attended a team meeting in Bill Fouts' office to discuss the stop order and plan a possible course of action to get the Attitude Reference System production line moving again. In attendance at the meeting were Bill Fouts, Bill Paxton, Bill Fordiani, Phil Woltman, Milt Swan, and me. We reviewed the quality assurance problems that American Wianko was having and decided that Stan Remick should set up a meeting with Bill Fouts' team and American Wianko. Later that morning, after Phil Woltman had talked to Stan, Phil told Fouts' team that Stan had requested that the meeting be held at 2:00 p.m. At 2:00, Phil Woltman, Milt Swan, Bill McClousky of S&ID's Quality Assurance Department, and I met with Stan Remick. Attendees discussed American Wianko's problems so that everyone would have a clear understanding of what we were facing. The S&ID team learned that the quality control problems American Wianko was having with the Attitude Reference System were discovered during a physical inspection of the product. Stan agreed to call American Wianko and set up another meeting, which was ultimately set for 9:00 a.m. on January 25.

During the January 25 meeting, the discussion centered on the four problem areas that American Wianko had listed in the message sent to Stan Remick. I had gone to see Bill Fouts to relay the message that American Wianko's Quality Control Department personnel had solutions for all four problem areas, but some of the solutions required approval of waivers, which had to be approved by S&ID and NASA. Three action items resulted during the meeting and were assigned to Phil Woltman for resolution. As a result of the action items that Phil had to work off and the possible waivers that American Wianko wanted to request for delivery of the Attitude Reference System, Phil and Stan Remick set up a meeting scheduled for January 27. In attendance at this meeting were Ford Miller, the NASA Resident Apollo Systems Project Office project engineer for AFRM 009; Nick Panza and Jack Davidson from the Resident Apollo Systems Project Office Quality Assurance Organization; Ron Black, the American Wianko Attitude Reference System project engineer; Armand Da Vanzo from the Apollo Electrical Engineering Department; Stan Remick; Bill Paxton; and me. Of the four quality assurance problems that had precipitated the Attitude Reference System stop order, two of the problems had been resolved internally at American Wianko, but two needed further work and approval from NASA and S&ID. Attendees at the meeting agreed on a plan to resolve the two remaining problems. A couple of days later, these issues were resolved by the attendees at Stan Remick's meeting on January 27, and the manufacturing line stop order was resolved by American Wianko.

Later, on January 29, 1965, I was talking to Larry Hogan, who was acting systems engineer for Automated Systems, and Larry mentioned that by February 1 or 2 he would have a definite American Wianko Attitude Reference System delivery date from Jackson Cole, which ideally would be by the end of February. By the end of January, it seemed as though the Attitude Reference System problems were over, but that was not to be.

On February 12, Stan Remick called Phil Woltman, who told me that American Wianko was going to wire S&ID the exact weight of the Attitude Reference System to be delivered to S&ID. At this same time, Stan dropped another bombshell on us: Stan told Phil that the Attitude Reference System being prepared for shipment for S&ID to use in the Quality Test Program had two printed circuit electronic boards with scratches on the printed circuit tracks, and American Wianko wanted to present them to S&ID for disposition. On February 15, 1965, American Wianko personnel were at the S&ID facility in Downey, California, with the two printed circuit boards that had the deep scratches. S&ID needed to inspect the boards to determine whether the boards were acceptable for use in the Attitude Reference System Quality Test. Our quality control lab personnel inspected the boards with a 40X microscope and determined that the printed circuit boards would be okay to use on Quality Test Attitude Reference System units. We filled out a Material Request for Disposition form that was presented to Bob Antletz, which he signed to approve the two boards to be used on Quality Test Attitude Reference System units.

Project Engineer Ron Black of American Wianko called me on February 26 to inform our team that the Attitude Reference System had completed its acceptance test without a problem and its delivery to S&ID was imminent. I advised Ron that Larry Hogan would have a purchase order sent to Stan Remick to direct him to purchase the Attitude Reference System. By this time, on March 2, because of all the problems that had been solved, Fouts' AFRM 009 team was under the impression that all the Attitude Reference System problems had been dealt with and the system delivery to S&ID would take place that day. But the idea that the Attitude Reference System problems were over was merely wishful thinking. The last thing that American Wianko needed to do, as directed by the acceptance test procedure, before the Attitude Reference System could be delivered to S&ID was have their quality control inspector perform a physical inspection of the system. In doing so, the inspector discovered that the frame of the Attitude Reference System did not meet the required physical dimensions as specified by S&ID documents. The solution to this problem was to initiate a Material Request for Disposition form for approval of the nonconformance of the Attitude Reference System dimensions. This was done and the form was approved, allowing the Attitude Reference System to be delivered to S&ID on March 2, 1965.

Other things were happening on the AFRM 009 project while we on Fouts' team were busy trying to get the Attitude Reference System stop order at American Wianko lifted on January 19. At this time, I had assigned Gene Rucker to support the Bread Board Control Programmer lab tests at Autonetics. After completion of the Bread Board CP tests, Autonetics was going to deliver the Bread Board CP to us in Downey so that we could run our tests with the Control Programmer connected to the Block I Stabilization and Control System as it would be connected in AFRM 009. The Bread Board CP configuration met the design requirements of the Rev B CP Procurement Specification with the new mission and the NASA-requested design changes except that it contained no redundant components, nor any high-reliability components, and was not packaged in a pretty production style, but it did perform all the latest AFRM 009 mission flyable functions. Because I had been assigned as lead engineer for the Control Programmer design, it was my responsibility to know exactly what was happening in every facet of the Control Programmer design, development, and test at S&ID and at Autonetics because I was called to go to Autonetics many times to sign documents or provide written directions. I communicated with Gene Rucker and the Autonetics engineers daily while I was working to resolve the Attitude Reference System stop-order problem with Phil Woltman. In this way, I received all good and bad news regarding the progress of the Control Programmer program. I needed to keep on top of everything because I was periodically called in by upper management to give a status report concerning the progress of the project.

As I indicated earlier, after the bread board was tested successfully at Autonetics, it was going to be delivered to S&ID so that my team (in conjunction with a lab team) could implement our own tests. After delivery of the Bread Board Control Programmer to Murphy's Lab, the plan was for Gene Rucker, Don Farmer, Thom Brown, and me to test the bread board to verify its agreement with the Procurement Specification before we turned it over to the Murphy Lab team, at which time they would run tests as a system, with the Control Programmer and the Stabilization and Control System functioning together as they would in the AFRM 009 mission flight. My team would act in a supporting role during the Murphy Lab systems tests. These lab tests would be the first time that the Control Programmer would be interconnected to the Block I Stabilization and Control System. For this reason, it was considered a big Apollo milestone.

On January 19, 1965, while Fouts' team was working on the Attitude Reference System panic, Gene Rucker reported to me that the Bread Board Control Programmer testing at Autonetics was coming along. Gene said that the test was in progress and, so far, they had encountered problems with the test panels, but the problems were being resolved as they went along. As soon as Gene and I were off the phone, I called Sam Rotuna at Autonetics to verify that the Bread Board Control Programmer was going to be delivered on January 25 to us at Murphy's Lab in Downey. I had to know this because John Shriver and Bud Weir, who were responsible for conducting the Control Programmer tests in Murphy's Lab, had to assure me that they would have all the necessary cables and test panels to hook up the Control Programmer to the Stabilization and Control System in time for the January 25 test. Sam said that they were still planning to deliver the Bread Board Control Programmer to Murphy's Lab on January 25. This news prompted me to phone John Shriver to inquire about the status of the rack being assembled for the Control Programmer and Stabilization and Control System, as well as the Control Programmer tester panel being built. John verified that he had a full set of connectors for the Control Programmer tester panel. The switches on the tester panel would be labeled with switch numbers, and John and Bud Weir would be responsible for connecting the Control Programmer patch panel to the Stabilization and Control System patch panel per print.

On January 20, 1965, Gene Rucker, still at Autonetics, had an update for me on the progress of the tests on the Bread Board Control Programmer; he said that the first run of the Control Programmer Normal Timer and the Control Programmer Abort Timer had been completed, and one problem had been encountered. The pitch torquer, the circuit responsible for commanding the Stabilization and Control System to fire the Command Module reaction control jets to orient the Command Module for Earth reentry from orbit, was blowing fuses, indicating that the circuit was drawing too much electrical current. The morning had been spent troubleshooting the pitch torquer circuit for faults, but none had been found. The plan was to replace the pitch torquer circuit fuse with a fuse rated at a higher current and reapply power to the Bread Board Control Programmer after lunch. After the telephone call with Gene, and during a call with Bud Weir at Murphy's Lab, I discussed the schedule for building all the panels and cables necessary to interconnect the Control Programmer with the Stabilization and Control System and the schedule for checking out all the panels and cables to my satisfaction. Bud provided the start and completion dates for each panel and cable, along with test start dates for their checkout and checkout completion dates. That afternoon, I sat down with Bill Fouts (bypassing Bill Paxton, of course) to review the schedule I had received from Bud, and Bill agreed that the schedule would meet the AFRM 009 requirements. Because of Bill Paxton's absence from the meeting, I was certain that Bill Fouts was aware that I was bypassing my supervisor.

Nothing on the program could be left to chance, so on that same date, in a phone conversation with Gene Karze and Sam Rotuna at Autonetics, I received confirmation that the delivery date for the Bread Board Control Programmer would be on or before January 22. I gave Gene and Sam the exact address and building location of Murphy's Lab in Downey so there would be no screwups on delivery. I immediately called Roy Murphy to let him know that the Bread Board Control Programmer was being delivered to his lab on January 22 by Autonetics and to be on the lookout for it. Later in the afternoon, Gene Rucker called me to report that testing of the Bread Board Control Programmer had been successfully completed and that delivery to us would indeed occur on January 22, but we would be getting no electronic circuit schematics with it. I found this news unacceptable, so I immediately went to see Bill Fouts directly, bypassing Bill Paxton again. Bill suggested that I call Autonetics and ask them whether they could send us a person who was familiar with the Control Programmer circuits to help the test team. That was exactly what I was hoping to hear. Sam Rotuna took my call, and I asked him whether Autonetics could send us a technician to help with the Control Programmer testing for two or three weeks, or maybe Autonetics could send us a set of Bread Board Control Programmer schematics instead; Sam said that he would call back the following day to relay the decision as to what Autonetics could do for us.

The following day, Gene Rucker was back from Autonetics, and Sam Rotuna called me and said that Autonetics was going to provide us with electronic schematics for the Bread Board Control Programmer. On January 22, Gene was at Murphy's Lab, and he called me to verify that Autonetics had delivered the Bread Board Control Programmer. Bud Weir and John Shriver realized sometime during the day that they needed extra cable connector pins and did not have them. I contacted Sam at Autonetics to see whether we could get some connector pins, and I was told that if Gene were at

A problem arose during Autonetics prototype Control Programmer tests. Look at the bottom row of six black diodes—the third diode from the left has a metal tab that has no wire attached to it, and slightly to the right and above the tab is a wire that was improperly soldered to the metal tab, now loose and floating (Courtesy of the Boeing Company).

A relay (an electronically activated switch) that was removed from the prototype Control Programmer box by Autonetics shown upside down with a row of three pins above and a row of four pins below. The next picture shows why the relay was removed (Courtesy of the Boeing Company).

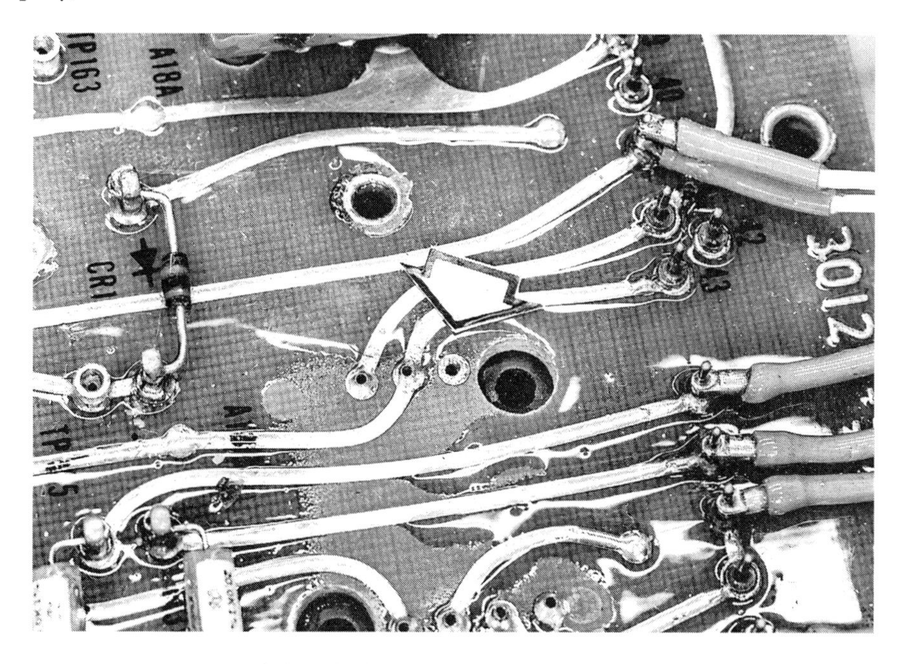

The arrow points to a row of three holes with solder splash on the left side. Below the three holes is another row of four holes. The relay in the previous picture was mounted on the board from the opposite side (as shown in next picture), with the relay pins matching the holes. The solder splash caused a short that resulted in the failure of prototype Control Programmer tests at Autonetics. The problem was a quality control issue corrected by procedures at the manufacturing plant to preclude a reoccurrence (Courtesy of the Boeing Company).

Autonetics on Monday, January 25, he could pick up the connector pins that morning. Gene was indeed at Autonetics on Monday morning to pick up the connector pins and a set of Bread Board Control Programmer schematics that Sam had promised.

From January 25 to February 1, Gene supported the continuity testing of the Bread Board Control Programmer in Murphy's Lab that was conducted per a procedure written by Mike Falco of Larry Hogan's unit. The test uncovered no problems, so on the afternoon of February 1 I went to Murphy's Lab to give the okay to apply power to the Control Programmer. When power was applied, two problems were encountered. The problems, it was discovered, were due to a missing ground wire in the cables that Bud and John had built. After the cable had been repaired by installing the missing wire, power was again applied to the Control Programmer, and the Lockheed timer motors began to purr like a beautiful, well-oiled machine. Even though the Control Programmer was made of metal, numerous electronic components, and many feet of electrical wire, I felt an intimate relationship to it because I knew that it was the "Thing" that was going to orchestrate the flight of AFRM 009, and I had helped design it. While Gene, Bud and John were preparing to start the Control Programmer tests, the electricity was flowing through the Lockheed timer motors and the electronic circuits in the Control Programmer, and the boxes of the system began to warm up. As I put my hand on one of the Lockheed timers and felt the warmth and the slight vibration from the Lockheed timer motors, I got this weird feeling that the Control Programmer almost felt like a living organism. Right there and then, I knew that we were moving one step closer to a shot at the moon by the end of the decade.

The RCC box of the prototype Control Programmer at Autonetics shows numerous relays mounted on the top side of the board. Relays are identified by three dots on the top side of the relays and four dots on bottom side (Courtesy of the Boeing Company).

On returning to the office that afternoon, I was talking to Carl Conrad about the Bread Board Control Programmer being in Murphy's Lab, and I told him that the test team had applied power to it. Carl jumped out of his chair toward me, looking straight at me with fire in his eyes, and as he was coming out of his chair he yelled, "You can't apply power to my Control Programmer until I say so." Because Carl had a lot to do with the functional design of the Control Programmer and had a personal attachment to the Lockheed timer, he believed that he owned the Control

Another wire that came off a diode metal tab that appears close to the center of the picture during an Autonetics test of the prototype Control Programmer (Courtesy of the Boeing Company).

Programmer. I looked straight at Carl and told him, "It isn't your Control Programmer, Carl; it's my Control Programmer because I'm responsible for the hardware, not you, and anyway I've already applied power to it." I let Carl know that the Control Programmer had not burned up when power was applied to it and that all was okay, which seemed to settle him down. The Control Programmer and Stabilization and Control System testing continued until March 5, with Bud Weir, John Shriver, and Cruz Mora (another Eagle Knight) from Murphy's Lab running all the tests, supported by Gene Rucker, Don Farmer, and Thom Brown. I had never met Cruz Mora until that day in Murphy's Lab. Cruz told me that he had transferred from the Los Angeles Airplane Division, where he was an engineer working on the B70 airplane.

Cruz was the sixth Hispanic engineer I had met at North American Aviation. When I started working in corporate America in 1954, there was no diversity in the professional ranks. At that time the only professional Hispanic I met was an engineer by the name of Quiroz back when I worked for Hughes Aircraft Company in Los Angeles. In 1960, when I came to work at North American Aviation Missile Division on the Hound Dog cruise missile, the only two Hispanic engineers I met were Paul Garcia and Frank Vigil, both of whom probably got their degrees after World War II on the GI Bill. As time went on, while working on the Apollo program, I started to meet more Hispanic engineers; no doubt most got their degrees on the GI Bill after the Korean War, as I did.

There were many problems during the testing period, too many to include them all here, but some of them are worth mentioning.

On February 3, 1965, the Bread Board Control Programmer test team stopped testing at 6:00 p.m. The ground commands in the Radio Command Controller unit had all been checked, and the team found that two fuses were blown. Gene Rucker contacted Autonetics to try to obtain fuse replacements, and the team had stopped and would continue testing the following day. On February 4, the Bread Board Control Programmer test team ran into trouble again, as they discovered a diode that had failed in the open condition in the Service Propulsion System engine thrust-on circuit. The team spent about two hours troubleshooting the problem and found that a diode was open. The problem was repaired, and the testing continued. On Friday, February 5, Gene Karze of Autonetics came to Bill Paxton's group area, hand carrying four seven-amp fuses for the Radio Command Controller unit that Gene Rucker had requested from Autonetics. Mike Falco hand carried the fuses to Gene Rucker and Don Farmer in Murphy's Lab.

That afternoon, the Control Programmer–Stabilization and Control System interface tests were started but had to be halted because water-glycol coolant was leaking from the Stabilization and Control System cold plate. The Stabilization and Control System had to be cooled to maintain the temperature of all its boxes. The method of cooling the boxes involved mounting each one on a specially designed cooling plate that had water-glycol flowing through it. Since the cooling plate was being repaired, no Control Programmer–Stabilization and Control System interface tests could be run until the following week. On Monday, February 8, the cooling plate was ready and the system was reassembled. The Control Programmer–Stabilization and Control System interface tests were restarted but had to be stopped once more because smoke was emanating from the Stabilization and Control System. The problem was resolved at 5:30 p.m., and the test team decided to resume testing the following day.

After completing a successful test that was witnessed by engineer Gene Rucker, this first prototype Control Programmer was used to verify the new mission with the Block I Stabilization and Control System in the Downey lab. The Radio Command Controller is in the background, the Lockheed timer in the middle, and then the Automatic Command Controller (Courtesy of the Boeing Company).

At this point, the Control Programmer test was now one and a half weeks behind

The AFRM 009 CP Lockheed timer with the Veeder-Root counters exposed (Courtesy of the Boeing Company).

Another view of the Lockheed timer Veeder-Root counters (Courtesy of the Boeing Company).

A third view of the Lockheed timer section. The drive motor was internal to this section (Courtesy of the Boeing Company).

The Lockheed timer section, looking at the back end with a 6-inch ruler to illustrate the size of the device (Courtesy of the Boeing Company).

Part of the equipment used to test the Block I Stabilization and Control System and prototype Control Programmer with the new mission (Courtesy of the Boeing Company).

The Block I Stabilization and Control System cable connections to the prototype Control Programmer and the lab test equipment (Courtesy of the Boeing Company).

schedule. When you are the lead engineer on a project, informing upper management that you are behind schedule does not make you popular. I pushed the guys in Murphy's Lab to hurry and make up some lost time and kept reassuring management that we would soon be back on schedule.

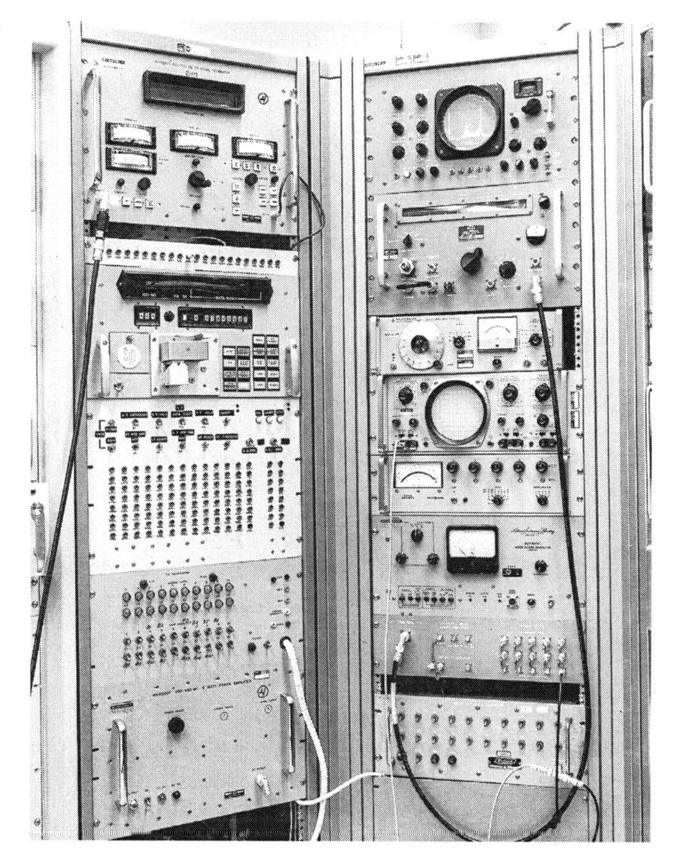

The equipment designed by Bud Weir to test the prototype Control Programmer with the Block I Stabilization and Control System in the lab (Courtesy of the Boeing Company).

Another big problem surfaced on February 9, and it took all day to locate its source and fix the issue. The Control Programmer–Stabilization and Control System interface tests were designed to verify that the Control Programmer could indeed operate normally when connected to the Stabilization and Control System. When power was applied that day, all hell broke loose. Instead of the system being in a quiescent condition, every Reaction Control System jet in the Command Module began to fire. The firing of the jets was manifested by Reaction Control System jet lights coming on and off on one of the test equipment control panels. The problem was fixed at the end of the day, and Gene Rucker, Cruz Mora, Bud Weir, Mike Falco, and Bell decided to work late to see whether they could make up some of the lost days. They had to stop testing at 10:30 p.m. because the nitrogen gas used to pressurize the glycol cooling system for the Stabilization and Control System had been depleted.

The original schedule (as given by me to Bill Fouts) called for all Bread Board Control Programmer testing to be completed by February 15; instead, on that date the team was just finishing the Radio Command Controller–Ground Control Communications System tests, which put us about eighteen days behind schedule. The test verifying that the Control Programmer could command the Stabilization and Control System to fire the Reaction Control System jets in proper order was run successfully on February 18, and on February 22 Gene Rucker brought the company photographer to the lab to take pictures of the Bread Board Control Programmer and the Stabilization and Control System in their test configuration. The next set of tests to take place was the Control Programmer–Communications interface tests, to be followed by Control Programmer braking tests. The interface tests were completed on March 2, and the braking

The cables connecting the Block I Stabilization and Control System to the prototype Control Programmer, as well as to other systems required for the lab test (Courtesy of the Boeing Company).

More cables connecting the Block I Stabilization and Control System to the prototype Control Programmer and test equipment (Courtesy of the Boeing Company).

tests were run successfully on March 5, which concluded the Bread Board Control Programmer tests.

So now, with the successful conclusion of the tests in Murphy's Lab confirming that the CP and the Block I Stabilization and Control System could work well during the AFRM 009 mission, it seemed that the program was one step closer to getting AFRM 009 ready for flight. Also, the March 26, 1965, issue of *Skywriter* (the company newspaper) reported that the Apollo heat shield had been structurally qualified for flight through extensive rigorous structural testing. All the program needed now was to have an AFRM 009 delivered out of manufacturing to the testing department in Building 290, and a Control Programmer delivered by Autonetics, and the tests on AFRM 009 could begin so the company could deliver AFRM 009 to NASA, but that step would come later.

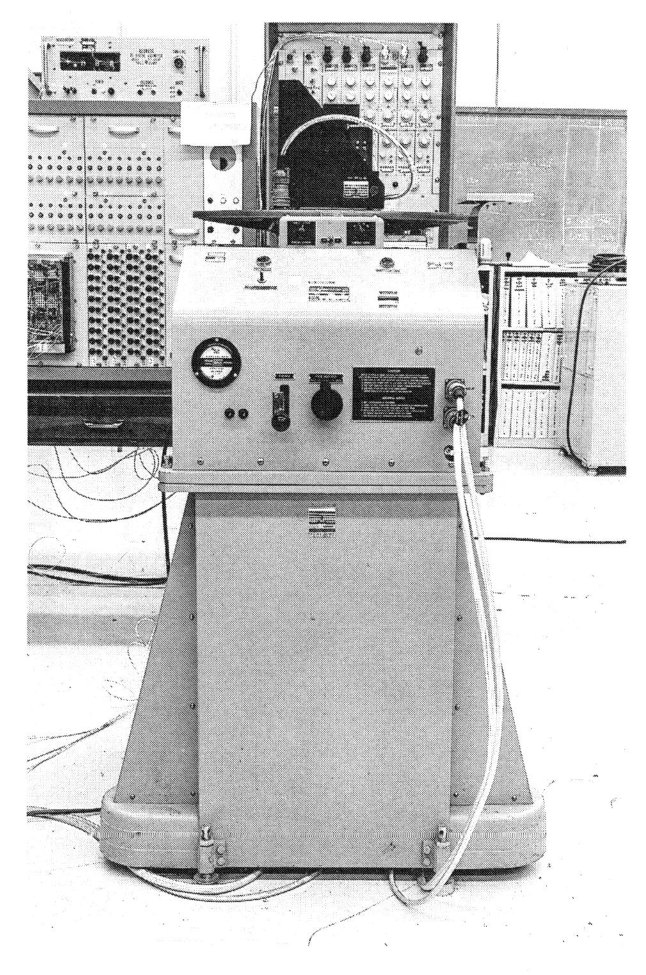

The rate table where the Stabilization and Control System BMAGs were mounted during the first prototype Control Programmer lab test (Courtesy of the Boeing Company).

One of the objectives that the AFRM 009 Control Programmer had to meet was the environmental qualification tests that were mentioned earlier. Don Farmer was the engineer whom I had assigned to oversee this part of the program, which began to take shape on January 26, 1965. The Control Programmer Qualification Tests requirements were delineated in a Space and Information Systems Division Report dated April 1, 1965. This document listed the tests that would start at Autonetics on July 21, 1965, and be completed on November 15, 1965. When working on any Control Programmer design problem or ongoing test, common practice for our team was that we all had primary responsibilities, but when we had free time each one of us was required to help one of the other two so that the work would get done expeditiously. The Autonetics engineer responsible for the Control Programmer Qualification Tests was Bill Knox. On January 26, 1965, Don Farmer met with Bill and his team at Autonetics, and an agreement was reached on the final form for the qualification test plan, electromagnetic interference plan and the acceptance test procedure. The qualification test of the Control Programmer was rescheduled to take place from May 15, 1965, to July 15, 1965, agreed to by

Don and Bill, so Bill and his team were working hard to get all test plans, test procedures, mechanical fixtures and equipment ready. Chuck Markley, who was the mechanical engineer in the Automated Systems Design unit, was given the task of reviewing all the mechanical fixtures that Bill had designed and built in order to run the Control Programmer Qualification Test.

The qualification test of any Apollo system determined in part the reliability of a system; for this reason, any changes made to the Control Programmer Qualification Test had to be coordinated with Bob Simpson of our Reliability group (headed by Ed Wheelock at the time). For example, on February 3, Bill Knox called and requested a change in the S&ID-required sequence of the environments to which the Control Programmer would be exposed during the qualification test. Don Farmer met with Bob Simpson to submit Bill's request for his review. Don and I had discussed the Control Programmer Qualification Test program in detail for many days and had identified areas where possible problems could occur, so on that same day in February, I called Sam Rotuna at Autonetics and scheduled a meeting to take place in Downey on February 8 at 10:00 a.m. to discuss the Control Programmer Qualification Test program.

On Monday, February 8, Don Farmer, Bob Simpson and I met with Sam Rotuna and Bill Knox at S&ID in Downey as scheduled. The five of us discussed the Control Programmer Qualification Test procedure in detail. After the meeting, I met with Bill Fordiani to discuss some changes that S&ID wanted to make to the Control Programmer Qualification Test program. It was decided that a request for a Budgetary and Cost Estimate would be submitted to Autonetics, and Don Farmer would write the letter of request to the IDWA group so they could communicate it formally to Autonetics. By March 18, Don had received and reviewed the Control Programmer Qualification Test procedure from Autonetics and was ready to discuss his comments on the procedure with Bill Knox. Don and I placed a call to Sam Rotuna to see whether a meeting was necessary to pass Don's comments on to Autonetics. When Don read his comments to Sam over the telephone, Sam agreed that a meeting was not required; instead, Sam would forward the comments to Bill Knox. While Don and I were still on the phone with Sam, Sam suggested that May 1 be tentatively be set for the Production Control Programmer design review. Don and I agreed with Sam on this date and told him that we would propose an agenda to them for the design review.

8

The Baseline Control Programmer Design

In later meetings (which will be described in detail later), agreement was reached regarding a Control Programmer baseline on which to base the changes to be made by the new mission. As mentioned previously, the first design review on the Control Programmer took place at Autonetics on February 12, 1965, to establish a Control Programmer baseline configuration. There is a reason why Autonetics was given the Rev B of the Control Programmer Procurement Specification with the new mission requirements and the NASA-requested changes in 1964 and why John Rowe of Autonetics released the schematic that had the Rev B Control Programmer Design on December 10, 1964 (as documented in chapter 6), and why my Control Programmer Design Team waited until February 12, 1965, to address the baseline Control Programmer design. We in the AFRM 009 Program needed a Production Control Programmer design as soon as possible to be able to support the AFRM 009 schedule delivery to NASA, and Autonetics likewise needed the design quickly so they could order electronic components and special mechanical and electromechanical parts to support their production schedule. There was no urgency for having a baseline Control Programmer design on paper early in the program because it would not be needed until the Production Control Programmer was delivered to S&ID; then Autonetics could make a cost comparison for all the changes made to the Control Programmer to arrive at the Production Control Programmer against the baseline Control Programmer to project a firm cost.

The February 12, 1965, design review to establish a baseline Control Programmer was followed by a design review of the Production Control Programmer that would eventually be installed in AFRM 009. The Control Programmer design review that took place on February 12 to establish the Control Programmer baseline configuration was held at Autonetics; that date had been agreed to on January 27 during a telephone conversation between Supervisor Bob Fettes, Sam Rotuna of Autonetics, and me. We agreed at that time that Autonetics would deliver Control Programmer schematic drawings to us on February 2 and that Bob Fettes and his team would write a design review agenda. On February 2, Sam Rotuna called Downey to inform me that Mitch Wieser was hand carrying Control Programmer schematics for the Control Programmer design review. The Control Programmer schematics that Mitch had brought to Downey were reviewed, and it was agreed that they were sufficient to support the review ten days later. Don Farmer, Gene Rucker, and I read the design review agenda that Mitch had brought along with the schematics, and it was agreed to be discussed later. On the following day, Sam Rotuna called again to let me know that Thom Brown was hand carrying some more

Control Programmer schematics to Downey. Thom brought the Control Programmer schematics in to work on February 4, and on that same day John Rowe came to Downey with another bundle of Control Programmer schematics with changes and updates to replace some of the schematic sheets that Autonetics had previously provided to my team.

My team members and I, along with some of the members of Carl Conrad's team, met in a conference room and spent time from February 4 to February 9 reviewing the schematics to verify the design of the Control Programmer, being careful to note all the design deficiencies so they could be discussed at the design review. Bill Paxton and I discussed the Autonetics-proposed design review agenda at length and found some issues and made some comments to it that had to get resolved before February 12. On February 9, I took part in a conference call between Sam Rotuna, John Rowe, and Bill Paxton, at which time the agenda comments and issues were settled.

In support of the design review for the baseline Control Programmer, a letter was issued on February 10 by John Beauregard (who worked for Larry Hogan) that described the intended design of the Control Programmer. It was a revision of John's previous letter, originally dated September 16, 1964, and contained enclosures of the updated Control Programmer drawn by John and enclosures for the Stabilization and Control System pitch, roll and yaw channels drawn by Tim Hickey. Because John worked for Larry Hogan, the letter described the official updated Control Programmer to be discussed during our design review. I met with John Beauregard, Lead Engineer Carl Conrad, and Don Farmer on Thursday, February 11, to determine Downey's stand and position on the Control Programmer design issues to make sure that Carl's team and my team were in agreement for the design review the following day. This design review meeting was crucial in determining the Control Programmer baseline because John Beauregard's letter, coming from Larry Hogan's unit, spelled out the requirements that had been translated into the Control Programmer Procurement Specification by Bill Paxton's design unit, and we needed to ensure that both units were on the same wavelength regarding the design of the Control Programmer.

The first Control Programmer design review to establish a baseline configuration went off as scheduled on Friday, February 12, at Autonetics. Design issues were discussed, and the result was that there were numerous Control Programmer design deficiencies that would have to be corrected or eliminated. Also discussed were additional Control Programmer design changes that S&ID wanted to make that would no doubt result in design ROMs from Autonetics. A ROM was a "rough order of magnitude" estimate given to us by Autonetics that was somewhat more accurate than a SWAG. All attendees agreed that I was to meet with Autonetics Project Office and program management personnel to resolve all the baseline Control Programmer design issues and all design changes that required ROMs.

On Wednesday, after spending a couple of days reviewing all the potential Control Programmer design changes for the baseline agreed to the previous Friday, I met with Project Engineer Jerry Paccassi and Program Manager Don McLean at Autonetics to devise a plan of action for resolving all the Control Programmer design issues that had been discussed during the design review on February 12. The three of us agreed to delete four items that would have required Control Programmer redesign ROMs. We also agreed that Autonetics would take care of all design deficiencies required to make the Control Programmer baseline, and I would live at Autonetics with Lead Designer

John Rowe until all deficiencies were resolved by correction or elimination to our satisfaction. I returned to Downey at 5:30 p.m. that same day with a list that had been generated at the Autonetics meeting of mandatory design changes to be made to the Control Programmer that would resolve all the design review issues on the baseline. On the following day, Thursday, February 18, I returned to Autonetics to meet with John Rowe to re-mechanize the Control Programmer to incorporate the mandatory changes that had been agreed to with Jerry and Don. At the end of the day, John and I had redlined about 90 percent of the six mandatory changes that had been identified during the design review into a clean and up-to-date baseline Control Programmer schematic.

Being committed to live at Autonetics until the Control Programmer was completely re-mechanized, I returned to Autonetics the next day, at which time John and I completed the redlines. For this task, I drew on my three years of teaching radar electronic circuits in the Air Force, the electronic circuit design courses I took at USC and the two years as a test equipment design engineer at Hughes, plus the four years of work on the design of the Hound Dog Cruise Missile Flight Control System. The revised schematic was then used by Autonetics to redesign the Control Programmer prototype hardware and to revise all other necessary Control Programmer drawings to give us an acceptable baseline. On Monday, February 22, 1965, I spent the day informing all engineers, designers, and management in a conference room in Downey about how John and I had re-mechanized the Control Programmer to clear all the design deficiencies resulting from the review. After many discussions with the various managers and engineers, and providing rationale as to why John and I had re-mechanized the Control Programmer the way we did, everyone agreed that the re-mechanization was acceptable and that we now had a baseline Control Programmer design from which Autonetics could give Bill Fouts' department a firm cost proposal on the design changes resulting from the Rev A and Rev B of the Control Programmer Procurement Specification. Convincing everyone in Downey to accept the new Control Programmer mechanization to which John and I had agreed is a good example of how the spoken word, the written word and salesmanship come into play for an engineer to succeed in any marketplace.

Having finished re-mechanizing the Control Programmer per the six mandatory design review changes for a baseline and having received agreement from all concerned in Downey, it was now time to work on our requested changes that would require ROMs from Autonetics, so on Tuesday, February 23, I returned to spend the day with John Rowe again. The problem that had to be solved was one that John and I called the 470-second time delay. We eventually arrived at a solution by adding some electronic components and rewiring a section of the Control Programmer. On Wednesday, February 24, 1965, I explained the 470-second-delay mechanization at a meeting with Bill Fouts, Larry Hogan, Bill Paxton, Carl Conrad and his team, as well as my team, in Downey. At the end of the day, I was ready with comments, possible improvements and agreements to the design change for the next meeting with John Rowe. On Thursday afternoon, John and I worked on preparing the ROM for the 470-second time delay to be submitted to Bill Fouts' department. John and I worked out an acceptable mechanization that included suggestions and improvements that I had brought from Downey, and we agreed to meet on Friday to thoroughly review the work and make sure we had not goofed up the Control Programmer. The 470-second time delay final design change mechanization that John and I had reached was agreeable to management and members of both teams in Downey on Friday, so that afternoon I headed for Autonetics for the final review. No problems were found with the 470-second time delay ROM during our final review, and the ROM was officially sent by Autonetics to us in Downey.

This decision ended my live-in period at Autonetics and freed me to attend to other problems. My thoughts were that now we would soon have a baseline Control Programmer design that met the agreement of S&ID and Autonetics, but the ordeal was not over yet. The Downey IDWA group still had to give Autonetics the formal go-ahead on the six mandatory changes that John Rowe and I had made to the Control Programmer to achieve the baseline design, and that authorization would have to be channeled through Bill Paxton for an unknown excruciating delay for me, as well as Autonetics, because the Autonetics folks would be calling me for a status on the formal approval of the design changes. Following the Paxton delay, Autonetics would have to provide Control Programmer delivery dates; until then, we would not know whether they could support our required dates for AFRM 009.

On Friday, February 26, 1965, I was on the telephone with Jerry Paccassi to discuss the Control Programmer mandatory design changes that John Rowe and I had identified to clear the baseline Control Programmer design review deficiencies. Without these changes, the Control Programmer would not be able to complete the initial original ARM 009 mission and there would not be a valid baseline Control Programmer. Jerry told me that with these changes to the Production Control Programmer, Autonetics could support delivery of the first prototype Control Programmer on April 15, 1965, and the second prototype Control Programmer on May 3, 1965. Jerry told me that he had passed the mandatory changes to manufacturing and requested that they inform him immediately about what they needed from engineering to support delivery of the Production Control Programmer by the required date. Jerry called me back on Monday, March 1, to verify that May 15 was a valid delivery date for the Production Control Programmer. The dates that Jerry quoted were in line with the AFRM 009 schedule. The Production Control Programmer was not really a production system as we know production products. Because this system had Apollo-qualified parts and was certified for flight, we called it the Production Control Programmer to distinguish between it and the prototype Control Programmers that intentionally were not qualified for flight on an Apollo capsule. Remember that Apollo was a research program; there was nothing production about it. No one had ever been to the moon before, so everything about the program was research and development, especially AFRM 009, because it would be the first Apollo capsule qualified for space flight to go into space.[1]

On Thursday, March 4, 1965, the six mandatory changes that John Rowe and I had redlined on the Control Programmer schematic were at the stage of implementation by Autonetics. A letter of authorization for the changes would follow. For the letter of authorization, on Wednesday, March 10, Gene Rucker from my team wrote the letter from Bill Fouts to J.F. Olsen, directing Autonetics to make the six mandatory changes to the Control Programmer. On this same date, the letter was in Bill Paxton's office for review. On March 17—one week later—Gene's letter was finally completed and signed. It was another unnecessary delay, and for me it was a week of agony because Autonetics management kept calling me asking for the status of the letter. All I could tell them was that it was out of my hands and to call Bill Paxton, who now had the written copy for his review. Taking eight hours to review a letter was probably okay, but 40 hours was ridiculous, incompetent and irresponsible. The problem was that Bill was completely rewriting the letter instead of just editing it. The letter from Bill Fouts to J.F. Olsen went out on the same day that Gene initialed it, and the Control Programmer changes were then officially approved. We in Downey now had a Control Programmer baseline configuration, and the Downey AFRM 009 teams had achieved their first AFRM 009 major milestone.

During the week of February 26, 1965, astronauts Dick Gordon and Dave Scott were visiting the S&ID facility in Downey to review a model of the probe that would be used to dock the Lunar Excursion Module to the Apollo capsule when the two astronauts rendezvoused with the capsule to dock on their return from the moon. There is a photograph of the two astronauts accompanied by three NASA personnel, with the five of them reviewing a model of the probe, on page 3 of the March 5, 1965, issue of the *Skywriter.* With the increased astronaut traffic in the Downey facility, it seemed to me like the Apollo program was getting closer and closer to a manned mission. But before the Apollo program could even consider a manned flight, my team had to regain some of the schedule days we had lost and get AFRM 009 off the ground.

Sometime in 1965, Phil Woltman was fired. Phil had been hired by Bill Paxton at my recommendation to be on the Control Programmer team to replace Z. Stanley Kleczko to handle the AFRM 009 backup Attitude Reference System. Phil lived on Turnbull Canyon Road in the Puente Hills between Whittier and Hacienda Heights, California, on an acre plot where he and his wife were building an adobe brick house. Phil commuted from there to work, driving down the hill on windy Turnbull Canyon Road into Whittier, from which he would head for Downey. One morning Bill Paxton's group received a telephone call and was told that Phil was in the hospital due to an accident. Apparently, Phil had gone over the cliff into the canyon in his pickup truck as he was heading to work and ended up at the bottom of the canyon. We were later informed that one of the suspected causes behind the accident was that Phil had purposely driven off the road in an attempted suicide. At the time, this was not a big shock to members of our team because prior to the accident Phil had been behaving strangely. He began by frequently coming in late to work and leaving early, and then he started to disappear for long periods of time during the day. When all this trouble started to become obvious, Bill Paxton talked to Phil, and Phil promised that he would shape up. Bill even mentioned all this to me and said that I had given him a bum steer by recommending Phil as a good guy to hire. I told Bill that when I knew Phil on the Hound Dog cruise missile program, he was a good worker; I had recommended him in good faith, and I was not aware of any previous strange behavior. Anyway, it was later disclosed that Phil had an alcohol problem. After his accident, Phil's behavior continued to deteriorate, becoming even more bizarre, and Bill talked to him again about coming in late and leaving early every day. Phil's reply was that he was twice as good as anyone in the group and because of that he did a day's work in half a day, so he was justified in working only half a day for a full day's pay. Phil had no intention of changing his behavior, so Bill had no recourse but to fire him. Before firing Phil, Bill met with the Human Resources Personnel Department, and a justifiable case file that spanned a period of months, detailing Phil's misbehavior, was prepared by Bill. On a specified date that Bill Paxton had agreed to with the Human Resources Department, Phil was given his pink slip, relieved of his badge and ID card, and escorted out the gate by Bill.

Because several major design changes were made to the Control Programmer in 1965, there were multiple design reviews held between us and Autonetics. The next Control Programmer design review on the agenda was for the Production Control Programmer, and preparations for that step started with a phone call to me from Sam Rotuna on

March 18, 1965. Sam said that Autonetics would have the Production Control Programmer design completed by the end of April, so he tentatively suggested a May 1 date for the design review. The date was agreed to, and I agreed to prepare an agenda for the review. On March 19, John Rowe released the C Revision to the Control Programmer schematic, which was another milestone for the AFRM 009 Control Programmer program. I proceeded to put together a list of suggested topics for discussion during the May 1 Production Control Programmer design review, which I circulated to Thom Brown, Gene Rucker, Don Farmer, and Steve Parrish on Monday, March 22, for comments and suggestions. Several days later, the copy of the list of topics had circulated with initials next to each person's name but with no comments or suggestions. My list was completely ignored, and my comment at the time was "I guess I get no inputs from these Cats."

On Wednesday, March 24, 1965, at the management meeting with Autonetics, the Control Programmer Qualification Test was discussed. The qualification testing of the Control Programmer went on as scheduled at Autonetics starting in May and ending in July. Don Farmer and Gene Rucker supported the environmental test according to the plan that had originally been rejected by Bill Paxton and that I had presented to Dave Levine (see chapter 5). The Control Programmer Qualification Tests took place at Autonetics on a twenty-four-hour schedule; Don and Gene split the days, with each one working a twelve-hour shift, with me on call during the twenty-four-hour day (which included Saturdays and Sundays). There were plenty of challenges, and many times I had to go to Autonetics to help solve one of the many problems or to write a letter of direction to Autonetics to get the qualification test program back on track. In fact, there were so many problems for a while that Don called me at home one night, and, boy, was he mad. He called me to ask whether it was okay to issue a stop order to Autonetics; there were so many issues that night that Don wanted to stop the testing on the Control Programmer until Autonetics fixed everything. I advised Don that the Control Programmer team could not do that because, in the first place, if S&ID stopped Autonetics from testing the Control Programmer, it would cause a schedule delay, and the only person who had the authority to delay AFRM 009 was Charley Feltz. In the second place, Autonetics was probably already behind schedule, and if S&ID stopped the Control Programmer Qualification Test, it would give Autonetics the opportunity to claim that S&ID's stop order had caused the delay. For S&ID, that would be a double whammy; they would blame us for the stop order delay and for their own schedule delay. I sensed that Don was really upset that evening, and so, to calm him down, I decided to get out of bed and drive to Autonetics to talk some sense into him. Fortunately, despite all the problems, the Qualification Test of the Control Programmer was completed in 1965 in time to support the AFRM 009 program.

During the 1:00 p.m. management meeting at Autonetics on Wednesday, March 24, the date of May 1 was confirmed as the date for the Production Control Programmer design review. I spent the rest of the week putting together an agenda for the design review, and on Monday, April 5, I distributed the tentative agenda to Larry Hogan, Bill Paxton, Chuck Markley, Bill Fouts, Steve Parrish, Thom Brown, Gene Rucker, and Don Farmer. I even gave an information-only copy to Mitch Wieser.

To get the Control Programmer ready for the launch of AFRM 009, numerous other things took place in the AFRM 009 Control Programmer Project from early January 1965 to at least April 1965. There were many more changes made to the Control Programmer and additional problems that needed to be solved before a Production

Control Programmer was delivered to us by Autonetics for AFRM 009, but they will not be detailed here. The week of April 9, 1965, was a momentous one for the Control Programmer and for the Apollo program. During this week, NASA accepted the airframe structure for the AFRM 009. This meant that the structure of AFRM 009 was going into Building 290 in Downey, where the various subsystems were going to be installed and checked out under test conditions for proper verification of operation, which included the Control Programmer. As appropriate for the delivery of AFRM 009 to NASA, this same week, there was a photograph of the AFRM 009 structure in Building 290 with S&ID's Don Taylor; F.K. Weidler, NASA Western Operations Office at Downey; and J.A. Davidson, MSC office, discussing NASA's acceptance of AFRM 009.[2] Another edition of *Skywriter* has a photograph showing Test Project Engineer Ted Clauss briefing company president J.L. Atwood; Charlie Feltz, Apollo assistant program manager; and S&ID president Harrison Storms on the checkout program for AFRM 009 during their visit to Building 290. This all indicated that Apollo AFRM 009 was on its way to meet its launch date and check off one more of many milestones for the Apollo program.[3]

The thing that made us feel more confident that the Apollo program was getting closer to meeting the moon launch date was that a cadre of astronauts was in Downey the week of April 9, 1965, to receive a five-day course of Apollo systems. Those astronauts in attendance were Charles Bassett, Dick Gordon, Don Eisele, Walt Cunningham, Alan Bean, Dave Scott, Michael Collins, Rusty Schweickart, Gene Cernan and Roger Chaffee. All of us on the Apollo program were aware that the astronauts were in Downey because some of us invariably ran into them walking down the halls during that week or saw them in the Apollo Mockup in Building 1 at our Downey facility.[4]

9

Preparing AFRM 009 for Launch

For AFRM 009, 1965 started off with a bang when the airframe was delivered to the Apollo Systems Integration and Checkout Facility by the manufacturing department. This milestone in the AFRM 009 project marked the beginning of testing the systems in the airframe in preparation for delivery to the NASA Manned Spacecraft Center in Houston to be assembled with all the rockets at the Kennedy Space Center, where it would undergo pre-launch tests. The S&ID company weekly newspaper ran a photo in its January 8, 1965, issue showing Harrison Storms, the division president, and Dale Myers, division vice president and Apollo program manager, with manufacturing personnel at the delivery of AFRM 009 to the Apollo Systems Integration and Checkout Facility.

The testing of AFRM 009 systems would begin even though the Control Programmer and its Attitude Reference System had not yet been delivered by Autonetics and American Wianko, respectively. Paul Rupert, who was either a graduate of Caltech or was attending Caltech for his master's degree, was the Control Programmer Test responsible engineer in the Apollo Systems Integration and Checkout Facility in Building 290, and Thom Brown in Bill Paxton's design unit was the responsible engineer supporting Paul during the Control Programmer and Attitude Reference System tests.

Thom was a graduate of the University of Kansas and had done a tremendous amount of work to help get AFRM 009 ready for Paul to start testing. Thom coordinated with Paul and other AFRM 009 Test Team Organization personnel almost daily for test requirements at the Downey test facility in Building 290 and for test interface requirements at Cape Canaveral. Thom also released some internal letters to make design changes to AFRM 009 to facilitate its testing. Some letters involved range safety issues, and others concerned test equipment. If you recall from chapter 2, Thom Brown was the engineer whom I befriended during my first days on the unmanned Apollo capsules and whom I took to Autonetics so he could meet the Control Programmer team. Following Thom's first trip to Autonetics, he assumed the task of having Autonetics provide the Control Programmer with the circuitry so that the Control Programmer could be tested when installed in AFRM 009. As it turned out, Thom, as I had hoped, became a main player in the Control Programmer design and was perfect for the job because he had transferred from field engineering, in which system testing is their main task.

Because the Control Programmer was a one-of-a-kind system for AFRM 009 only and standard tools could not be adapted to the installation or retraction of the Control Programmer hold-down bolts, Thom and I had to devise and build a special tool to insert and extract the bolts when installing or removing the Control Programmer from the mounting platform in the Command Module. Without this tool, Building 290 technicians would have experienced serious problems with installing the CP on the AFRM 009 platform and would have been in worse trouble when the time came to remove the Control Programmer boxes for repair.

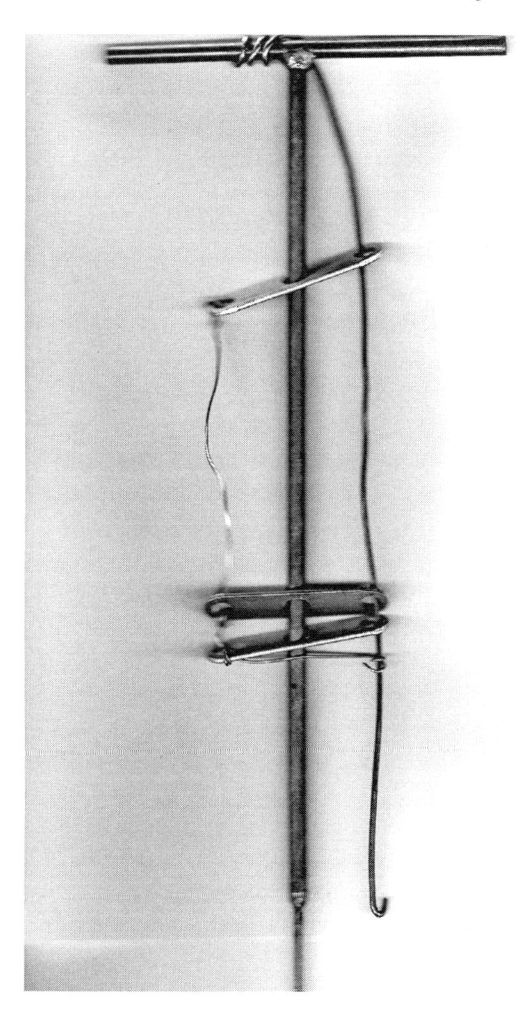

This is the special tool made by engineers Thom Brown and Jim De La Rosa to install and unbolt the Control Programmer boxes from the platform in AFRM 009 (Courtesy of the Boeing Company).

Autonetics delivered the Production Control Programmer for AFRM 009 on April 9, 1965, and was immediately installed into AFRM 009. The May 16, 1965, issue of the *Skywriter* newspaper ran a picture of Ted Clauss, AFRM 009 Test Project engineer, briefing the AFRM 009 checkout procedure in Building 290 to Lee Atwood, company CEO; Charlie Feltz, division vice president in charge of AFRM 009; and Harrison Storms, division president.

Because of the growth of the company, around the week of May 21, 1965, it was announced that all the engineering groups that were in Building 6 in Downey were going to be moved. The company had acquired a building in Downey on the north side of Lakewood Boulevard at the intersection of Bellflower Boulevard, which was designated as Building 318. The relocation of personnel, desks, filing cabinets and equipment would start on May 24 and continue through June 12. Bill Paxton's design unit was scheduled to move on Saturday, May 29. The move was a good thing for those of us going to Building 318 because the building had a large parking lot behind it and finding a parking spot was not a problem. By contrast, the facility where Building 6 was located was huge, and even though there was plenty of space for parking, if one came in late to work for any reason or was returning from a subcontractor facility late in the day, the only available parking slots were located quite a walk from any of the buildings.

On Sunday, May 23, 1965, the *Los Angeles Times* ran several articles that stated emphatically that the United States would keep its moon date and described AFRMs 009 and 011 and Spacecraft 012. AFRM 011 would be the unmanned airframe to follow

AFRM 009, and Spacecraft 012 would be the first manned Apollo spacecraft to orbit Earth—it was generally referred to as Apollo 1 and would be a Block I Command Module like AFRMs 009, 011, 017 and 020.[1]

There was one occurrence at S&ID sometime in 1965 that is worth mentioning here: For quite a while at S&ID, it was customary to separate all excess material that was not needed (such as desks, tools, tables, wood, and any other similar materials), declare them surplus material and sell them to the public at the S&ID surplus store. The store gave first choice to employees on Fridays to purchase items; on Saturdays, as usual, the surplus store would open to the public. Our family would visit the surplus store periodically to see whether we could find something of value that we needed. One Monday morning, when we returned to work, all employees were asked whether they had visited the surplus store the previous Friday or Saturday; those who answered in the affirmative were given a form to fill out, listing all the items they had purchased. At the time, the questions that were asked sounded sort of weird, but later the significance became clear. The S&ID Apollo capsule manufacturing department was told, sometime during the AFRM 009 manufacturing process, that because AFRM 009 was going to be unmanned and would not carry any astronauts, some of the capsule's specific control panels that contained switches normally controlled by astronauts were not needed because their functions were going to be performed by the Control Programmer; thus, those panels had to be removed from the Command Module. What manufacturing had neglected to mention was that the control panels removed from AFRM 009 must be saved for use with one of the subsequent Command Modules. Instead, those control panels were declared "excess material" and sent to the surplus store, where they were sold. All that week, inquiries were made of all employees, and by the end of the week everyone knew that the AFRM 009 control panels had been sold at the surplus store. Those employees who had been at the store that Saturday were even asked whether they had observed any person buying control panels that had electrical switches on the front of them. Unfortunately, the AFRM 009 control panels were never found. Even at this late date, there may be a person (or persons) out there somewhere in possession of control panels from the Apollo Command Module AFRM 009.

On June 3, 1965, we received a firm cost proposal on the Control Programmer for the baseline configuration based on the original mission from Autonetics, which was promptly reviewed and submitted with comments to Tom Whaling of the IDWA group before the due date of June 8.

The June 4, 1965, issue of the company newspaper, the *Skywriter*, reported that during the next week there would be a joint NASA/S&ID review of Spacecraft 009 known as the Development Engineering Inspection (DEI). As reported in *Skywriter*, on Monday, June 7, 1965, the AFRM 009 DEI was held in the Apollo Mockup Area in Downey and continued through the week, ending on Thursday. The DEI was a very formal review, and the active participants were high-level managers from NASA and North American Aviation Space and Information Systems Division; the process included a review of the spacecraft, its systems and the ground support equipment required to completely check out the spacecraft systems while installed in ARFM 009 in Building 290. The objective of the DEI was to review the checkout philosophy for systems and the vehicle to the satisfaction of NASA. To give NASA a heads-up on the DEI, the week before the beginning of the inspection, two S&ID employees—Bob Milliken, who was manager of Apollo Ground Operations Requirements, and Al Dale, manager of Apollo

Checkout and Integration—made a briefing presentation to management at the Manned Spacecraft Center in Houston on the Apollo Spacecraft checkout logic and plans. Al had been the Hound Dog cruise missile project engineer from 1960 to 1964 when I was working on that program.

The final discussions during the Development Engineering Inspection were held between the NASA high-level managers, the North American Aviation high-level managers who were all designated as board members and North American Aviation lower-level engineering managers. Engineers, including lead engineers like me, were required to attend because we were responsible for the system that was to be installed in AFRM 009, and the higher-level managers needed us there in case a system detail question came up that only we could answer. We were a backup team, so to speak, and spoke only if the high-level manager requested that a backup team member speak. I was issued a pass for the DEI and attended the meeting as required.

On June 7, the DEI engineering participants who were going to conduct the actual review were divided into five working groups that included the North American Aviation Space and Information Systems Division lower-level engineering managers and their NASA counterparts. Heading the working groups were Joe Dyson, manager of Guidance and Control, who led the G&C Working Group (of which I was a member); Dr. J.I. Dodds, manager of Apollo Systems Dynamics, who led our Dynamics, Sequencing, Sensing and Recovery Working Group; Jim Johnson, manager of Apollo Command Module Structural Design, who led the Command Module Working Group; Nort Nelson, manager of Apollo Electrical Systems, who led the Electrical and Mission Support Working Group; and Bob Kurtz, manager of Telecommunications, who led the Telecommunications, Information and Acquisition Working Group.

The high-level managers for North American Aviation and members of the board were Dale Myers, vice president and Apollo program manager; Charlie Feltz, Apollo assistant program manager; Gary Osbon, Apollo chief engineer; Norm Ryker, director of Apollo Systems Engineering; Ray Pyle, chief project engineer; and Jim Pearce, director of Apollo Test and Operations. The high-level managers for NASA and members of the board were Dr. Joe Shea, manager of the Apollo Spacecraft Office; Chris Craft, assistant director for Flight Operations; F.J. Bailey, chief of the Flight Safety Office; Dr. Max Faget, assistant director for Engineering and Development; Deke Slayton, assistant director for Flight Crew Operations; and G.M. Preston, deputy director of Launch Operations, Kennedy Space Center.

The setup for members of the board at the Development Engineering Inspection in the Apollo Mockup Area was a row of conference tables at the front, where the high-level managers from NASA and North American Aviation sat; in front of the conference tables were chairs, where the individuals from the working groups sat, and behind them sat the audience. There were several microphones on the conference tables, as well as a roaming microphone with a long cord for the members of the working groups. The purpose of the microphones was to amplify the discussions for the audience and to record all the vocal proceedings. The NASA high-level managers asked the questions, and the North American Aviation working group leaders answered whatever questions they could; if they could not, then the question was referred to one of the lower-tier workers. Normally the answer given by the working group came from the working group leader, but occasionally, if they were unable to answer the question in full, they would seek help from the expert on the subject in the group. If the S&ID working group leader could not answer the question at all, he would allow the system expert to use the microphone to respond. The NASA or S&ID high-level manager had to speak into a microphone set up at the conference tables, identify himself and then ask the question. Likewise, the S&ID working group leader (or whoever he appointed to answer the question) had to identify himself when speaking into the roaming microphone.

The proceedings started on Monday, June 7, 1965, with the meeting breaking up into the five working groups on Monday and Tuesday to discuss the checkout philosophy for each system and the vehicle. During these two days, action items were developed for NASA and North American Aviation S&ID by each of the working groups. The working group meetings were less formal than when confronting the members of the board. On Wednesday, the status of the AFRM 009 program was briefed to all attendees by A.A. Tischler, manager of Apollo Development Control, and John Tropila, Apollo Ground Support Equipment project engineer. On Thursday, June 10, the action items that resulted from all the working group meetings were reviewed in front of the board and assigned to NASA or to North American Aviation personnel. In certain cases, North American Aviation S&ID could not provide NASA with an answer because S&ID was waiting for material, contractual direction, or money, so NASA had to accept an action item to go back to Houston and move someone off the dime to provide whatever it was that S&ID needed to close the action item.

The Development Engineering Inspection was only one example among many of this type of meeting with NASA that some engineers were required to support because, after the AFRM 009 Development Engineering Inspection, there were still four years left on the Apollo program before a mission to the moon was possible. Many more Apollo Command Modules remained to be delivered to NASA, which had to undergo similar lengthy procedures.

On June 22, 1965, through the IDWA group, we formally redirected Autonetics to change the Control Programmer mechanization to the new AFRM 009 mission and asked them to provide us with a budget and pricing on the new mission changes. Autonetics agreed, and Vaughn Slaughter informed me that they would have a firm cost proposal on the new design by July 1, 1965. At the next management meeting, Vaughn presented the Autonetics budget and pricing for the new mission changes. Their original estimate for the Control Programmer had been $4.5 million, but the budget and pricing for the new mission Control Programmer was now down to $3 million because of redundancy reductions, the reduction of Control Programmer systems and the reduction of the number of factory test equipment sets. The Control Programmer Automatic Command Controller needed to be re-mechanized, but the Radio Command Controller would need only minor changes. The budget and pricing did not include $750,000 charged against the minor IDWA for the original mission baseline design. The meeting discussion then turned to the details of the reduced number of Control Programmer systems to be delivered to the Space and Information Systems Division, as well as the reduced number of pieces of factory test equipment.

We and Autonetics had multiple meetings to define all of Autonetics' problems with the new mission redesign, and they in turn required certain things from us to support them in their efforts. My team worked on all the problems that were facing Autonetics. Most of the items that Autonetics had requested help on required letters of direction that had to be channeled through the IDWA group. The letters of direction to Autonetics were all written by Don Farmer, Gene Rucker, and me, and all were rewritten in their entirety by Bill Paxton over a period of weeks before they were finalized to be initialed by each one of us and signed for release by Bill. The final signed copies were sent to the IDWA group, with the usual long delay, for transmittal to Autonetics. Once again, the delay caused by Bill rewriting the letters was agonizing for me because Vaughn Slaughter or Bob Fettes would call continually, asking about the status of the letters of direction, and all I could tell them was that the letters were written and were in Bill's office for review, and if they wished to know more, they should talk to him. The letters of direction finally got to Autonetics, and at last, after several more meetings, we agreed on the work to be done to re-mechanize the Control Programmer to the new mission, the number of Control Programmers to be delivered and the amount of factory test equipment to be built.

Bill Paxton released the new organization chart by internal letter on July 2, 1965, which was initialed by Bill Fouts; the updated version was necessary because of the Guidance, Navigation and Control department reorganization, new personnel, and new responsibilities. However, the new organization chart still listed me as lead engineer on AFRM 009, with most Attitude Reference System work completed, and only Gene Rucker for Control Programmer development and Don Farmer on Control Programmer Support and Attitude Reference System development. Mitch Wieser was listed as Mission Control Programmer Design lead engineer, with Frank Chee in Mission Control Programmer Development and John Kelleher on Mission Control Programmer Support. Pat Erlich was now gone from the program, and his empire was split between Thom Brown and Chuck Markley. Thom had been elevated to lead engineer for Automated Systems Test, with Steve Parrish on Process Specifications and Operational Checkout Procedures and J.E.G. Strong on Test Equipment. Chuck was promoted to lead engineer for Design Support and had the largest block on the chart, with Tim Hickey on Drafting, Bob Novak on Interface Compatibility, Milt Swan on Documentation, Schedules, Configuration Management, and Bill Vance on Automation Requirements.

As the Apollo program started to gain momentum, there were many activities taking place. In July 1965, Robert E. Greer, a 1939 graduate of West Point and a retired major general, was appointed assistant to Harrison Storms, S&ID division president. General Greer came to the position well qualified. Prior to his appointment, he had served in the Secretary of the Air Force Office, as deputy commander for Satellite Programs at the Los Angeles Air Force Station near LAX, as assistant chief of staff for Guided Missiles in Headquarters, USAF, and as chief of the Special Division in the Pentagon.[2]

The Mississippi test facility opened a manmade waterway to receive the first space age hardware: the Saturn S-II fit-up fixture, which had the same dimensions and weight of Bob Antletz's stove pipe Saturn II booster. The S-II fit-up fixture was used to check out test stands and other facilities and train crews at the Mississippi test facility. The test facility at the White Sands Missile Range in New Mexico, near Las Cruces, was the scene of additional activity, using the Little Joe booster rocket that was launching Apollo Boilerplates to test the development of the Launch Escape System rocket motor that would be used to take the Apollo capsule and its crew to safety in case something went wrong with the rocket boosters during the launch phase.[3] Avco Corporation of Wilmington, Massachusetts, was actively working on the Apollo heat shields so they would be able to withstand the 4,500 degrees Fahrenheit that the Apollo capsule would encounter for sixteen seconds during Earth reentry. The heat shield was ground tested under high heat conditions, but the MCP flights of AFRM 017 and AFRM 020 would certify the device.[4]

To those employees working on Apollo, all this ongoing development and testing provided impetus to move faster toward the moonshot goal.

At the management meeting on July 31, 1965, Vaughn Slaughter presented the schedule for the new mission changes to the Control Programmer System delivery based on the "A" revision of the Procurement Specification. Vaughn also alluded to some problems that Autonetics was facing, but he said these problems would not jeopardize the schedule. While Autonetics was working to deliver equipment to us, Gene Rucker, Don Farmer, and I were regularly visiting Autonetics.

Things on the Apollo program were not slowing down at all, but I figured that I needed to get away for a while, so I took a much-deserved vacation from August 23, 1965, to September 5, 1965. On September 9, a few days after I returned, John Rowe released Revision D of the Control Programmer Mechanization Schematic.

As mentioned earlier, Lead Engineer Thom Brown was responsible for supporting Paul Rupert in the AFRM 009 Control Programmer testing in Building 290 while AFRM 009 was undergoing test before it was delivered to NASA. Thom and I had an agreement that if Thom ever needed my help, Thom could call me anytime at work or at home. It was Thom's and my job to ensure that the Control Programmer worked as designed, and if it failed to perform as advertised, it was our job to find out where physically it had failed, fix it, and continue with the tests. Thom handled supporting Paul Rupert without my help throughout the AFRM 009 testing period except for one occasion. This incident took place on a Sunday evening around late September or early October. It was after dinner at our house when the phone rang. When I answered it, Thom explained to me that they were in trouble with the AFRM 009 Control Programmer. One of their tests was failing, and Thom and Paul's team had spent many hours troubleshooting the system on AFRM 009 and still could not figure out what was causing the problem. Thom said that they were stalled and had already lost many hours—a delay the program could hardly afford. He asked whether I could possibly come down to help them because I could no doubt bring some desperately needed fresh ideas. I lived about fifteen minutes from the plant, so I told my family that I would be back in a couple of hours.

I had figured that it would take a couple of hours of work because I was counting on doing the troubleshooting as it had been done on the Hound Dog cruise missiles, in which the engineer could talk directly to the technician, but I was mistaken. This was the NASA Apollo program, complete with high-reliability parts, strict documentation for traceability and all work done in an environmentally clean facility where white clean coveralls and head covers were required.

I had previously been issued a Spacecraft 009 Test Team badge by General Supervisor Paul Fields (the same Paul Fields I had met in Florida on the Hound Dog cruise missile program at Bill Harris' house the night I was invited to dinner along with Claire Harshbarger during the XGAM-77 Missile 029 jitter fix flight), so I had no problem entering Building 290 that evening. (Some words on the persons named above: When I was first hired at North American Aviation by then supervisor Bob Antletz to work on the design of the Hound Dog cruise missile in 1960, Bob's lead engineer was Claire Harshbarger. Bill Harris was an engineer on the test team at Eglin Air Force Base in Florida, where I spent three months in 1961 helping Bill test and launch Hound Dog cruise missiles, and later in 1963 I was there helping to launch Hound Dog cruise missiles with the fix for the missile jitter problem that could have resulted in contract cancellation by the Air Force; at that time, Claire was my supervisor. Paul Fields was a North American Aviation employee who was Bill Harris' friend.)

When I arrived and entered Building 290, I discovered that Thom, Paul, managers, and all other engineers were working in the control room, using headsets to communicate with the technicians who were doing the troubleshooting on AFRM 009 inside the clean room. At least I did not have to wear the white coveralls and head cover. However, the troubleshooting was something else. The rules had changed drastically from what I had been used to during my technician days at Hughes Aircraft Company and on my trips to Florida during the Hound Dog cruise missile program. All system troubleshooting had to be thought out methodically beforehand, written down on a trouble report and approved by a quality control inspector. Then the instructions had to be entered in the spacecraft logbook and approved by a quality control inspector. Each trouble report was limited to one step in the troubleshooting process. It took a long time to get enough data points to determine the source of the problem. By the time we really began to get something done, it was already around 9:00.

After I arrived, in talking to Thom and Paul, I could sense that they had exhausted all their ideas for finding the source of the problem. They told me that the test was failing at one of the test paragraphs of the procedure that could possibly be caused by a shorted component in the Automatic Control box in the CP. They willingly let me try my hand at leading the Control Programmer troubleshooting proceedings on AFRM 009. I started by looking at the Control Programmer Automatic Command Controller box schematic and linking the test failure to a circuit in the CP and then postulating in my head some possible sources of the problem and making sure it was not something that Thom and Paul had already tried. In devising the troubleshooting scheme, I started to tell Thom and Paul what the first steps would be in the process, but they stopped me cold and said that it could not be done that way—it had to be done the Apollo way. The troubleshooting was more formal than my previous experience, involved more people's approval for each step and consumed much more time; in addition, I was never allowed to talk directly to the technicians doing the troubleshooting. This procedure continued all night into the wee hours of the morning; I entered each troubleshooting step in a different trouble report each time until there was enough information and data to make a valid judgment on what had gone wrong. By this time, it was about 3:00 or 4:00 a.m. on Monday morning, but fortunately I had identified the source of the problem.

Using the Control Programmer Automatic Command Controller schematic and the troubleshooting data that I had accumulated, I methodically took Paul and Thom through the reasoning as to why there was a short to ground in a specific electronic circuit of the Automatic Command Controller. They both seemed like they believed my explanation, but then Thom said, "You mean to tell me that if I place a Mylar insulation sheet between the Automatic Command Controller box and the grounded platform it is mounted on, the problem will disappear?" For a moment, I had a slight doubt about my decision, but then I recovered and told Thom that what he had just said was true. I thought that would end the whole issue, but I was wrong.

Thom then took a blank problem report and started writing. His instructions were to unbolt the Control Programmer Automatic Command Controller box from the metal platform on which it was mounted with the special tool that we had constructed for that purpose and place a sheet of Mylar between the Automatic Command Controller and the platform. The next step Thom wrote down was to repeat the Control Programmer test that had been failing. Everyone working on AFRM 009 that night knew that if I was right, the test would not fail because the Mylar sheet would serve as an insulator, interrupting the ground path to the faulty short anomaly that existed in the Control Programmer Automatic Command Controller. As the technicians were performing all of Thom's steps in the problem report, my heart was pounding. What if I was wrong? "But no," I thought, "the reasoning for the source of the fault is right; there is no other explanation for the faulty behavior of the Automatic Command Controller unit during the test step when the failure occurred." By around 4:00 a.m., the rerun of the test requested by Thom was performed successfully, and as a result everyone was cheering. I told Paul and Thom to pull the Automatic Command Controller out of AFRM 009, as it would be sent to Autonetics for repair.

I called Don McLean, the Autonetics AFRM 009 program manager, at home; while I was on the phone with Don, informing him that the AFRM 009 Control Programmer was being returned to Autonetics for repair, Thom, Paul and other AFRM 009 managers were talking, and I could hear some of their conversation and heard words to the effect that the Control Programmer could not be removed from the spacecraft without Charlie Feltz's approval. Don told me that if we delivered the Automatic Command Controller to Autonetics that same day, it would be back at Downey within 24 hours. I told Don to send someone to pick up the Automatic Command Controller unit in Downey and gave Don instructions regarding what company gate the truck driver should come through to get to Building 290.

Looking through the hatch of AFRM 009, the large box in the background mounted on the platform is part of the Control Programmer (Courtesy of the Boeing Company).

As soon as I got off the phone with Don, Paul Rupert told me that the failed Automatic Command Controller could not be removed from AFRM 009 without Charlie Feltz's approval. So, I told him to call Charlie and tell him that the AFRM 009 Control Programmer had a problem and needed to be sent back to Autonetics for repair. Paul said that he could not call Charlie at home at 4:00 a.m. and wake him up. They were all afraid of Charlie, and no one there had the guts to call him so early. Paul told me that they would have to wait until Charlie came to work. I balked at this news

A North American Aviation technician on the Control Programmer platform working on AFRM 009 in Building 290 in Downey, California (Courtesy of the Boeing Company).

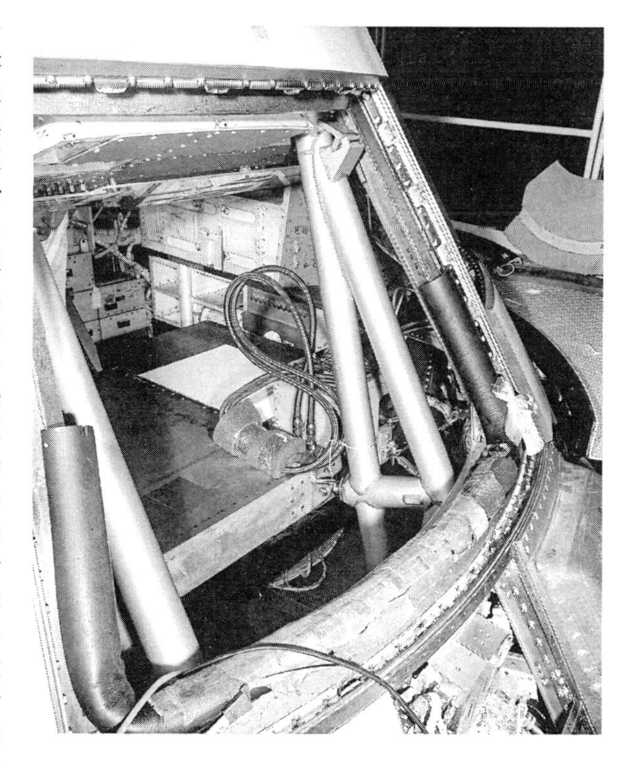

This is a view of the Control Programmer platform looking through AFRM 009's hatch in Building 290 (Courtesy of the Boeing Company).

and informed Paul that Autonetics would return the Automatic Command Controller to us in working condition within 24 hours of receiving it. If AFRM 009 Test Management waited for Charlie's approval (which probably would not happen until sometime late that afternoon), there would be a long delay, pushing back the AFRM 009 schedule even more. I told Paul that the sooner the Automatic Command Controller went out of there, the sooner it would be back. I also told him that not removing the Automatic Command Controller from the spacecraft now to send it to Autonetics for repair was irresponsible and that I was willing to take all the responsibility for its removal, and he could tell all that to Charlie. Paul reluctantly gave the okay for removing the Automatic Command Controller

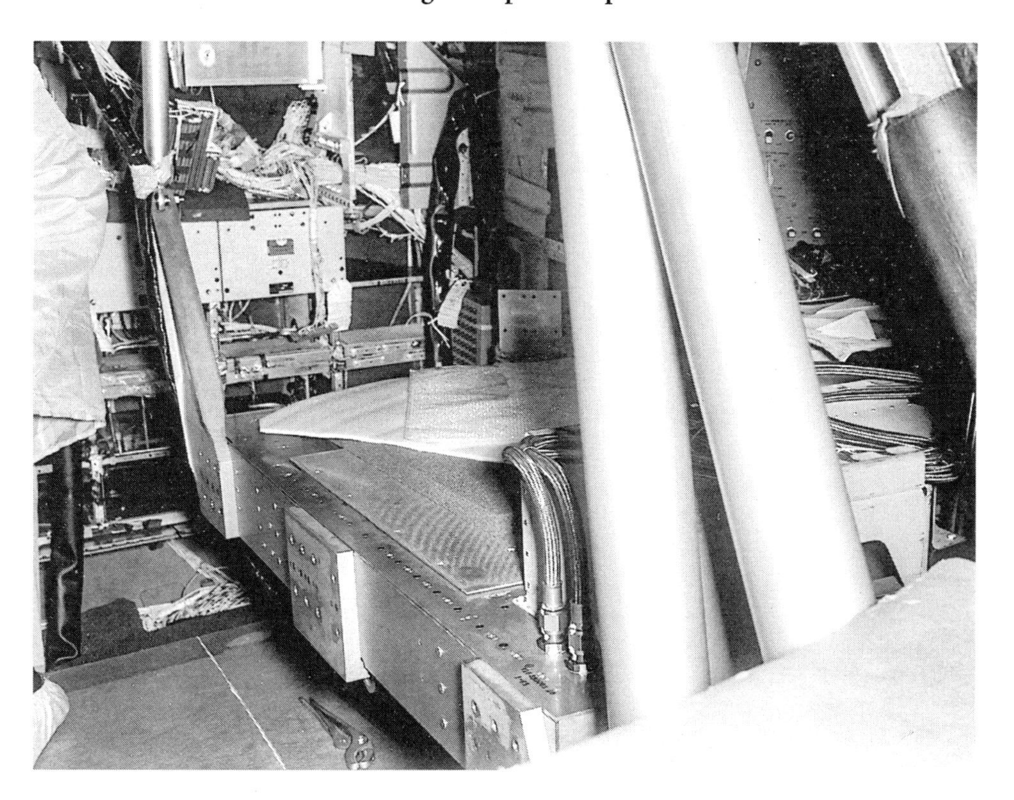

This is another view of the Control Programmer platform looking through AFRM 009's hatch (Courtesy of the Boeing Company).

but told me that Charlie Feltz was going to kill me. My reply to Paul was that Charlie was going to thank me for saving him valuable time.

The Autonetics truck driver arrived around 5:00 a.m. and departed with the AFRM 009 Automatic Command Controller. As I was thinking about it later, I started to wonder whether I should have listened to Paul Rupert and all his *pendejos* (dumb heads) and not have insisted that the CP Automatic Command Controller box be removed from AFRM 009 and sent to Autonetics for repair. Maybe I should have told Paul to call me later during the day whenever the failed CP box was ready to be sent back to Autonetics and let them eat whatever schedule delay the program suffered in addition to the 24 hours it took to repair the box due to their unwillingness to take responsibility. However, I had no regrets for what I did; as a matter of fact, I was proud of my actions because I could not have done any different.

That Monday morning at 7:00 a.m., after having worked through Sunday night, I was exhausted. I headed for the office to let management know what had occurred with the AFRM 009 Control Programmer. When I arrived, the only one in his office was Bill Paxton, so I told him all that had happened and that Bill Fouts would probably get a phone call from Charlie Feltz about it. When I was done talking to Paxton, I went to my desk to do some work before I headed home for some much-needed sleep. I kept checking to see whether Bill Paxton was going to give Bill Fouts the news. More employees had started to come in to work, and so I figured that by then Bill Fouts was probably already in his office. After about 15 or 20 minutes of no action by Bill Paxton, I headed for Bill Fouts' office, bypassing Paxton again, and found Fouts sitting at his

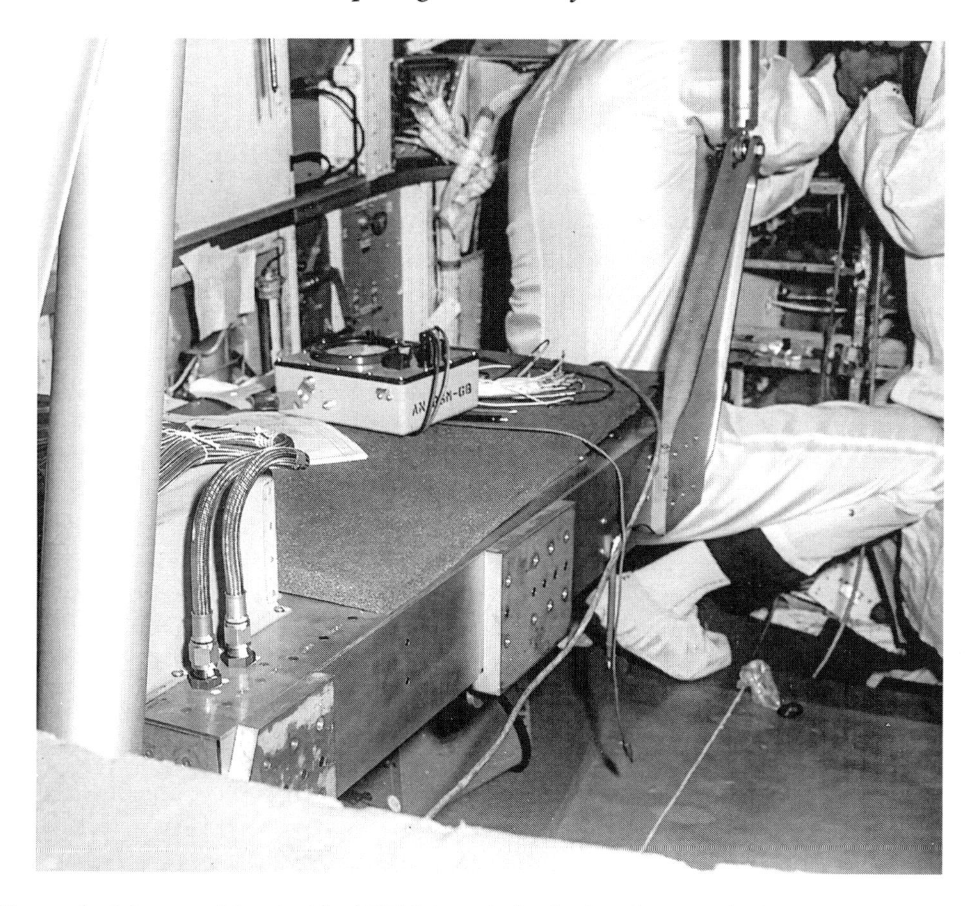

Two technicians working inside AFRM 009. A shock absorber attached to the Control Programmer platform can be seen adjacent to a technician's shoulder (Courtesy of the Boeing Company).

desk. I recounted the evening's events, starting with Thom Brown's phone call the previous evening to come in and help them and the resistance I had encountered when I wanted to pull the Automatic Command Controller from AFRM 009 to send it back to Autonetics. I explained that the only reason they pulled the Automatic Command Controller from AFRM 009 was that I had told them I would take full responsibility for pulling it and that they could tell that to Charlie Feltz. Because of that, I alerted Bill that he might get a phone call from Charlie. No sooner had I finished with the report than Bill's phone rang. He picked up the receiver, listened for a while and said, "Yes, I know, and he's standing right in front of me. The Automatic Command Controller will be back within 24 hours, as Jim promised." With that, Bill hung up the phone and thanked me, and I was out of there on my way home for some sleep. As I was walking out to my car, I was kicking myself for wasting my time in going into Bill Paxton's office that morning. What Bill Paxton did not understand was that no manager ever wanted to hear from a company vice president about a problem in any system under the manager's responsibility that he was embarrassingly unprepared to discuss.

I returned to the plant that afternoon after some sleep and some food. I contacted Sam Rotuna and Vaughn Slaughter and was told that they had found the electronic component causing the problem in the AFRM 009 Automatic Command Controller. A large power diode, as I had suspected, was being shorted to ground by some residual solder fragments that had inadvertently been left in the Automatic Command Controller box after some components had been soldered in place. The "fix" was to vacuum out the whole inside of the box, replace the shorted diode and revise the manufacturing procedures so that all Control Programmer electronic boxes would be vacuumed out after any component soldering had been performed just prior to installing the outside cover. I called Bob Simpson of Reliability Engineering and told him all the events regarding the CP fault in AFRM 009 and explained the solution for the problem; Bob agreed with the "fix," and that closed the issue.

But wait—the episode was not over yet. It turns out that when I returned to work that Monday afternoon after working 14 hours straight through Sunday night and Monday morning, the security guard at the gate where I entered the plant in the afternoon (after sleeping only a few hours) recorded my name as having come in late. Several days later, I was on report, and someone at the attendance administrative department demanded to know why I was being allowed to report late for work without being reprimanded or punished. Apparently I had broken a standing policy that anyone coming in late must be punished or reprimanded. My management had to write a letter and explain why I was late that day. When Bob Antletz became aware of my predicament, he said that when the new policy of reporting and punishing people for coming in late was first introduced at a management meeting, he had asked, "Will the employee be punished if he is absent from work?" He was told that no employee would be punished for being absent. Then Bob said, "When one of my employees calls and tells me he is going to be late, I am going to tell him to stay home and not come to work, because if he comes in late, he will be punished, but if he stays home, everything will be okay."

In the end, I never heard a word from the AFRM 009 Test Team about the work I had done for them; no doubt they had taken credit for my work and were lauded as heroes. As for me, well, I was the idiot who removed the Control Programmer box from AFRM 009 without Charlie Feltz's approval and had broken a company policy by coming in late the following day after working 14 hours straight.

During this period there were some similarly stupid policies in place that were eventually repealed. One mandated that when an employee needed to get a new pencil, they had to turn in the stub of the previously used pencil. Can you imagine giving adult graduate engineers kindergarten rules? This type of thing invariably happened when a person was promoted to a higher management position or when someone new was hired into a high management position from another company. They tended to bring their quirky ideas with them and imposed them on the troops. Managers who implemented these quirky policies did not last long, or else their quirky policies were soon repealed. How could the United States of America ever win the space race with idiotic policies like that?

10

The Launch of AFRM 009

The date for AFRM 009's launch was quickly approaching, and so everyone was looking at AFRM 009 with a critical eye. The engineers responsible for the guillotining and deadfacing of the umbilical cables during the separation of the booster rocket stages wanted to know whether their work affected the operation of the AFRM 009 Control Programmer and the AFRMs 011, 017 and 020 Mission Control Programmer. By this time in the program, I was lead engineer for the Control Programmer and the Mission Control Programmer, so I put my team to work doing a complete analysis of all Control Programmer and Mission Control Programmer wires that were cut during the guillotining process that occurred at booster separation. After a lengthy and thorough investigation, my team concluded that the Control Programmer and Mission Control Programmer operations were not affected by the guillotining of the cables. This issue was closed by an internal letter from Bill Paxton to R.H. Buch dated October 4, 1965, written by Bill Vance, asserting that umbilical cable guillotining and deadfacing during the booster separation phase of the flight did not affect the Control Programmer and Mission Control Programmer operation.

The last two tests that AFRM 009 underwent in Building 290 prior to delivery to NASA were the mission simulation and systems integration checkout, as was reported in the caption of a photo of AFRM 009 during testing.[1] With the successful completion of these two tests, the company was ready to deliver AFRM 009 to NASA and collect the hard-earned incentive dollars. The week of October 18, 1965, was devoted to the Customer Acceptance Readiness Review (CARR) for AFRM 009. The CARR was another of the formal reviews conducted with NASA that were mentioned in the previous chapter. Most lead engineers working on AFRM 009 were required to attend the CARR in case there were questions asked for which lead engineers needed to provide answers. Entry to the CARR meetings was restricted to individuals who had been issued the proper badge.[2] Most of the NASA management from Houston was in Downey for the series of meetings and presentations by S&ID to demonstrate the final result of all our hard work. S&ID management was successful in proving to NASA that AFRM 009 was ready for delivery, so on Wednesday, October 20, AFRM 009 was delivered to NASA in an informal ceremony held in Downey. Dr. Joseph Shea, Manned Spacecraft Center program manager in Houston, was there to accept delivery of AFRM 009 for NASA.[3]

AFRM 009 was now ready to go to the Kennedy Space Center, where it would be assembled with the Saturn S-IB booster in preparation for the booster's first flight and AFRM 009's historical mission. When launched, it would be a first step toward qualifying the Apollo heat shield for manned missions. This was only the first step because the reentry heating of AFRM 009's mission would not qualify the heat shield for the

5,000-degree-Fahrenheit heat of a moon reentry mission. Heat shield temperatures of U.S. manned flight spacecraft that went into space before AFRM 009 had been in the vicinity of 3,500 degrees Fahrenheit. However, AFRM 009's mission would qualify the heat shield for Spacecraft 012, the first Apollo Earth orbit mission carrying three astronauts. The way this would be accomplished was that, at an altitude of about 280 miles, AFRM 009 would push itself toward Earth using its 22,000-pound-thrust Service Propulsion System engine to a reentry velocity of about 18,750 miles per hour, which was approximately 10 percent higher than a normal reentry from an Earth orbit mission. The qualification of the Apollo heat shield for a moon reentry mission would be left to AFRMs 011, 017, and 020. Apollo capsule testing performed up to late 1965 had been of the ground static–type tests using Apollo Boilerplate airframe models. (Boilerplates were Apollo engineering test vehicles that simulated Apollo weight and shape.) Besides ushering the Apollo program from a model static-testing program to a dynamic flight-testing program using flight-rated spacecraft, AFRM 009 with its launch escape tower and the Service Module would also be the biggest spacecraft built to date. This accomplishment by AFRM 009 would be a first in the Apollo program and a great Apollo milestone.[4]

During this same time frame, the company was working other programs in parallel to Apollo. One such program mentioned before was the Paraglider program. The paraglider was a bat-like inflatable craft that could be used to return payloads safely from outer space or to return spent booster sections safely back to Earth for possible reuse, thereby saving money in the space program. Flight testing of this craft was ongoing at Edwards Air Force Base in the Mojave Desert. The program director was Bert Witte, and the company test pilots were Don McCuster and Jack Swigert. (Jack would later become a NASA astronaut and was pilot on the Apollo 13 mission. Apollo 13 was Apollo's near disaster in space and the subject of a movie by the same name and the spacecraft in which Jim Lovell uttered the famous words, "Houston, we've had a problem.") By October 29, 1965, AFRM 009 was already at the Kennedy Space Center undergoing final testing prior to its launch early in 1966.[5] The Command Module and the Service Module were stacked for integrated testing followed by stacking the Command Module–Service Module combination to the Saturn S-IB booster on the launch pad for booster compatibility tests, which were followed by an entire mission simulation run. On Friday, November 5, 1965, the Paraglider program came to a successful conclusion. Jack Swigert and Don McCuster flew the paraglider as it was dropped from an aircraft carrying a Gemini-type capsule to the Mojave Desert floor. Jack flew five missions, and Don flew seven. At the end of the program, Bert Witte, Jack Swigert, Don McCuster, Chief of Flight Control Claire Harshbarger and Paraglider Project Engineer Will Owens were out of a job. They were immediately absorbed by the Apollo program. Will Owens had been hired by Bob Antletz to work on the Hound Dog cruise missile program back in 1960 or 1961. While on the Apollo program, Will eventually became head of a directorate; Ralph Gomez, one of the Eagle Knights, mentioned to me that he worked in the Apollo Environmental Control System Department and that his director was Will Owens.[6]

The Saturn S-II booster (the stove pipe program that Bob Antletz had wanted to leave so badly) underwent torturous ultimate load tests on key components of up to 140 percent of design limit pressure on Saturday and Sunday, November 6 and 7, 1965, at Santa Susana, California, to certify their structural integrity for future manned moon missions.[7] Another milestone in preparation for the launch of AFRM 009 came when it was announced on November 19, 1965, that the Plastics Department had delivered the cover for AFRM 009 ahead of schedule.[8] The cover fits the spacecraft like a glove and is made of material that protects the command module from harmful elements in the launch environment during the boost launch flight phase. The cover, of course, is tapered like a tear drop to fit the command module like a glove. The cover hitches a ride as a protector of the airframe until it is jettisoned with the launch escape system tower later in the boost flight phase.

Around this same time, Vice President Bill Parker announced that M.A.G. Robinson (everyone called him Mag) had been appointed assistant program manager for program control on the Saturn S-II program. Around 1962 or 1963, my mentor and supervisor Ed Kelley and I had met Mag Robinson when Mag was one of the managers who had accompanied us to Emerson Electric in Silver Springs, Maryland, when a group of company specialists and managers were working there on the Hound Dog cruise missile program, resolving some radar altimeter problems. Mag was a West Point graduate and had come to work for the company in 1956 after he had separated from the service with the rank of lieutenant colonel.[9]

My team was devoting more time to the design of the Mission Control Programmer but was still working on some minor AFRM 009 problems because it was now late November, creeping into December 1965, and AFRM 009's launch date was just around the corner. On November 24, we received some test data on the Attitude Reference System from American Wianko, which proved that the Attitude Reference System was in good shape for AFRM 009's launch. The last update to the Control Programmer Procurement Specification—the D revision—was issued on December 15; two days later, I signed the checked block for the last release of the Control Programmer SCD, revision D.

The February 4, 1966, issue of the company newspaper, the *Skywriter*, had a nice long article about AFRM 009 and announced that the launch date had been scheduled for February 22, 1966. On February 18, AFRM 009 was already on the launch pad at the Kennedy Space Center, ready to go. About ten days prior to launch, the spacecraft and booster were put through a simulated countdown, which was followed by a flight readiness test and booster preparation (the final steps prior to launch). However, the February 25 issue of *Skywriter* let the whole company know that the AFRM 009 launch had been postponed due to bad weather in Florida. The last revision of the Control Programmer IDWA 6513 occurred in February 1966, eight days prior to the launch of AFRM 009. Before the document's release, I redlined the current revision of IDWA 6513 to incorporate the latest changes for the C revision on February 1. On that same day, Dick Gordon of the IDWA Group picked up the redlined copy of the IDWA. Dick returned a typed copy of it later for review and signatures. On February 18, the IDWA was signed, and on February 22 Bill Paxton met with Dick Gordon to turn over the signed IDWA.

Four days later, on February 26, 1966, AFRM 009 was launched from the Cape with the Control Programmer in control of the capsule. The vehicle performed a flawless mission. The Control Programmer orchestrated the graceful movements of AFRM 009 as it was propelled on its voyage in outer space to usher in the flight of Apollo. My Control Programmer team's hard work had paid off to help make history, and now we could work exclusively on the Mission Control Programmer for AFRMs 011, 017, and 020. The successful mission of AFRM 009, flying under the full control of the Control Programmer, was a tremendous milestone for the Apollo program not only as a NASA and company achievement but also because of the millions of incentive dollars that had been earned by our company as a reward for its scheduled delivery and its flight success.

There was great company jubilation because everyone knew that Apollo was on track for a moonshot in 1969. I was particularly proud because I knew that I and my team had contributed to its success and had helped to lay one of the first stepping-stones to a historical moon mission. The company magazine (*Skyline*) that was released after the launch of AFRM 009 had a feature article titled "Apollo Spacecraft 009 Returns to North American after Successful Mission." The article discussed how well the heat shield had performed during the AFRM 009 Earth reentry. To recount the history of space rocketry, the magazine also had an article about Dr. Robert Goddard, the father of space rocketry. The AFRM 009 media coverage did not stop there. The *Skywriter* newspaper announced that AFRM 009 was on its way home to Downey after launch on Saturday, February 26, 1966, on its 39-minute mission. Equally proud, the Autonetics *Skywriter* newspaper of the same date announced that the Control Programmer, Master Events Sequencer Controller, and the Service Module Jettison Controller (all manufactured by Autonetics) performed in an outstanding manner on AFRM 009 on February 26.[10]

The March 11 issue of the *Skywriter* newspaper had three pictures on page 3 of AFRM 009's recovery at sea by the carrier USS *Boxer* after its successful mission. On Saturday, March 12, and Sunday, March 13, AFRM 009 was on display at one of the company parking lots in Downey for viewing by all employees and their families. Needless to say, my family went to view AFRM 009. We had never seen a spacecraft when it had returned from outer space and experienced the effect of the reentry heat. The AFRM 009 heat shield was charred, and on the sides of the capsule you could see the blackened streaks where the reentry flames had scorched the ship. While viewing AFRM 009, I experienced an eerie and awesome feeling as I looked at the charred heat shield and the dark streaks that reached all the way to the capsule windows, and my immediate thought was that there were no astronauts to experience the flames and charring of the heat shield as reentry was happening—only the Control Programmer, which could not tell how it felt. According to the March 18 issue of the *Skywriter* newspaper, 10,600 people attended the weekend display of AFRM 009. The newspaper called the AFRM 009 a "Big Apollo Milestone" and said that the next Apollo mission would be Spacecraft 011 on a suborbital shot to be recovered in the Pacific Ocean.[11]

Of course, I already knew about AFRM 011 because I and the rest of my team were already working on the Mission Control Programmer for Spacecraft 011, 017 and 020. The Spacecraft 017 mission was to be the first flight of the Saturn V booster to qualify it for flight and to orbit the Earth and reenter Earth's atmosphere at a velocity that exceeded the entry velocity on a return trip from the moon. Spacecraft 020 would be the second flight of the Saturn V booster that would qualify it for moon manned missions and had a planned mission to head for the moon with only enough velocity to cold soak the spacecraft in deep space and then return to Earth. The AFRM 020 trajectory was to place the AFRM 020 payload in an Earth elliptical orbit and then fire one of its booster's engines to hurl it more than 13,000 miles in altitude, at which point it would then return to Earth and expose the heat shield to the moon reentry heat for a second time following the first heat shield exposure of AFRM 017. AFRM 020's mission was exceedingly critical because it would qualify the heat shield for moon reentry heat and verify the Earth entry guidance and navigation techniques as AFRM 020 came in through the narrowing corridor to thread the needle of Earth reentry.[12]

The big media coverage, of course, was the article about the AFRM 009 mission on two pages in Section A of the *Los Angeles Times* on Sunday, the day after the launch. The successful mission of AFRM 009 made the future mission of Apollo Spacecraft 012—the first Apollo manned capsule—seem more real. On February 11, 1966, Spacecraft 012 was in Building 290 in Downey undergoing systems installation and checkout. On that date, the storage space of Spacecraft 012 was studied by the astronauts. Those participating in the study were Alan Shepard, Walter Schirra, Virgil Grissom, Gordon Cooper, Russell Schweickart, Walter Cunningham, Donn Eisele, James McDivitt, Ed White, Frank Borman, David Scott, and Roger Chaffee. With the success of the AFRM 009 mission and the progress being made on Spacecraft 012 in the Apollo program, everyone on the engineering team felt confident that the moon landing would occur in 1969.[13]

It is probably now appropriate to mention some of the facilities that S&ID, in conjunction with NASA, was building during this period to support the Apollo moonshot in 1969 and some of the ongoing tests on various Apollo rocket components. Here are some things that were published in the 1966 *Skyline* magazine, volume 24, number 2: (1) After three years of construction and a cost of $300,000,000, the Mississippi test facility, where NASA would proof test the first- and second-stage boosters for the moonshot, was settling down to testing. (2) Another article described in detail the Saturn S-V and S-II boosters, the two heavy-lift boosters required for the Apollo capsule and Lunar Module to reach the moon in 1969. (3) The Apollo Altitude Chamber, where altitude tests of the Apollo Command Module would take place, was described in a third article, which included a photograph of Al Moyels, one of S&ID's research pilots, visiting it. (4) The fourth article described the technique that the Lunar Module pilot would use to rendezvous with the Apollo Command Module after lifting off from the moon. The rendezvous techniques had been developed and perfected on several of the Gemini orbital space flights by the astronauts.

With the "Mechanical Boy" and AFRM 009 now part of history, I and my team could devote all our time to the design and development of the "Mechanical Man" for AFRMs 011, 017 and 020, but could we keep up the fast pace for another two years without burning out?

11

Mission Control Programmer for AFRMs 011, 017 and 020

The way in which my team received the Mission Control Programmer assignment did not follow the proper protocol. This incident occurred sometime late 1965 at or near the end of the peak work on the Control Programmer for AFRM 009. As mentioned previously, the lead engineer for the Mission Control Programmer was Mitch Wieser, but apparently there was general dissatisfaction with his performance both at our department and at Autonetics. Sometime in late 1965, Bill Paxton called me into his office and said that, as of that day, I was lead engineer for the Mission Control Programmer, and I had to go tell Mitch that he was being relieved of his command. It was Bill's job as design unit supervisor (especially because it was Bill who had hired Mitch, not me) to tell Mitch of this decision, as any other supervisor at the time would have done, but Bill did not have the guts to go tell Mitch that he was being fired. Like everything else, if anything had to be done on this project, it was up to me to go give Mitch the news. I just went to Mitch and very diplomatically told him that Bill Paxton had told me that I was now the lead engineer for the Mission Control Programmer. Afterward, Mitch left the program (or maybe even the company, because he was never seen again). John Kelleher, who worked for Mitch, departed for another assignment, and Frank Chee, the other member of Mitch's team, went to work for Autonetics. From that day forward I was Mission Control Programmer lead engineer.

The missions of ARFMs 011, 017 and 020 were more significant to the Apollo program than AFRM 009's mission because they were designed to get Apollo one step closer to landing on the moon before the end of the decade. Please do not misunderstand me, because I am not attempting to minimize the importance of AFRM 009—it was after all, the first flyable Apollo capsule to go into space. However, the next three unmanned airframes to go out into space were going to test the Apollo capsule, including its installed systems, space guidance and celestial navigation techniques, in space in preparation for a manned moon mission. These three capsules were the last of the unmanned Apollo capsules to go into space, which would be followed by a manned mission. Like AFRM 009, they were also Block I capsules manufactured by S&ID in Downey, California. Because the Apollo capsule was to be the first manned spaceship to ever go away from Earth orbit and into space, there were many unknowns and dangers in sending the first manned Apollo capsule to the moon and many potential ways in which the astronauts would not be able to return to Earth and be lost forever in space or burn up during reentry.

The Guidance Navigation and Control system designed to take the Apollo spaceship to the moon and back to Earth had never been tested in a real situation in space with humans on board, and neither had the techniques for space guidance, navigation, and control. However, it is not true that guidance, navigation, and control techniques to the moon had never been proven before because there had been unmanned spaceships (such as Surveyor), but they were never required to return to Earth safely with humans on board. The guidance, navigation, and control techniques, as well as the exposure of the Apollo Command Module to the outer space environments, had to be tested in real life because proper operation of those systems in space was mandatory for the success of the Apollo moon mission, and what better way than to do this in an unmanned spacecraft?

One of the objectives of the last three unmanned Apollo capsules was to certify the heat shield system for a moon return mission. On the return trip from the moon, the Apollo capsule would be reentering Earth much faster than when reentering from orbits around the Earth, and so the reentry would be far hotter, requiring a better heat shield. This necessity meant that the Apollo heat shield had to be tested and qualified for moon reentry velocity if the Apollo crew were to survive Earth reentry, which was one of AFRM 020's objectives.

As mentioned previously, the window for Earth reentry on a return from the moon was very narrow, analogous to threading a needle. It was the job of the Guidance Navigation and Control System to keep the Apollo capsule inside this corridor, and so the MIT Draper Labs engineers who were responsible for developing this system to be installed in Ramon Alonso's computer had a tough job. All this was going to be verified by AFRM 020's mission, which would answer the question of whether these objectives were even feasible.

Like AFRM 009, AFRMs 011, 017 and 020 would be controlled by an electronic automated system that would automatically perform all the functions that an astronaut had to perform to fly the space capsules. The electronic automated system for these three spaceships was called the Mission Control Programmer. For short, we called it the MCP. We referred to it as MCP so often in our conversations that Milt Swan, who was on Bill Fouts' staff, began calling it "Emma-C-P" and eventually shortened the name to "Emma" instead of MCP. At the very beginning of this project, the Mission Control Programmer was called Mechanical Man, but that name quickly fell into disuse for the same reason that Mechanical Boy for AFRM 009 went by the wayside, and so MCP it was.

The missions for these three unmanned capsules were more complicated than the AFRM 009 mission, and so the Mission Control Programmer was more complex in design than the AFRM 009 Control Programmer. In its complexity, it tended to be more like a minicomputer rather than a simple motor timer system like the Control Programmer. It contained hardwired digital logic to allow it to make decisions depending on the information it was given. The technology of the MCP was not foreign to me because the one hundred hours of computer study I had received in 1958 when working for Hughes Aircraft Company included the study of the MA-1 Hughes digital computer, and its technology was the same type of hardwired computer logic. As mentioned before, AFRM 011 was to fly suborbital in a ballistic trajectory, AFRM 017 was to orbit the Earth and AFRM 020 was to go into deep space toward the moon and return to Earth, simulating a moon return trip; thus performing those various missions required a system approaching the complexity of a hardwired minicomputer.

In the AFRM 009 mission, the Control Programmer had a programmed timeline in a timer that performed all events and functions in elapsed time intervals. This design meant that AFRM 009 was really a dumb machine and was good enough for one type of mission only. In contrast, the next three unmanned spacecraft had three different and distinct missions. More electronic brains than a Control Programmer were needed for an Apollo capsule to be able to complete the diverse missions of AFRMs 011, 017 and 020. However, as complex as the Mission Control Programmer was, it could not fly a mission to land on the moon and return to Earth. To complete a moon landing mission before the end of the decade, Apollo needed a computer, a Guidance System, a Navigation System, an Environmental Control System, fuel cells for electrical power, and a Communication System that all worked properly, as well as a three-man crew and much more. As mentioned before, the Mission Control Programmer had complex electronic digital logic to be able to make decisions based on the various system inputs it received and to send out commands to the other systems so it could orchestrate the flights of the Command Modules for Spacecraft 011, 017 and 020. Because of its complexity and digital logic, one Mission Control Programmer could handle the three distinct missions of multiple spacecraft.

The Mission Control Programmer consisted of three electronic boxes: the Spacecraft Command Controller, the Ground Command Controller, and the Attitude and Deceleration Sensor. The function of the Spacecraft Command Controller was to make decisions with its digital logic based on input from various other systems to command the systems that controlled the spacecraft. The Ground Command Controller was intended to replace the function of the Spacecraft Command Controller in case it failed to work. The Ground Command Controller did this through its capability to receive commands from the ground up-data link to perform all Spacecraft Command Controller functions. The Ground Command Controller was used only if the Spacecraft Command Controller failed to operate properly. The failure of the Spacecraft Command Controller would be relayed to the ground by the Spacecraft Telemetry System, which then allowed the Ground Command

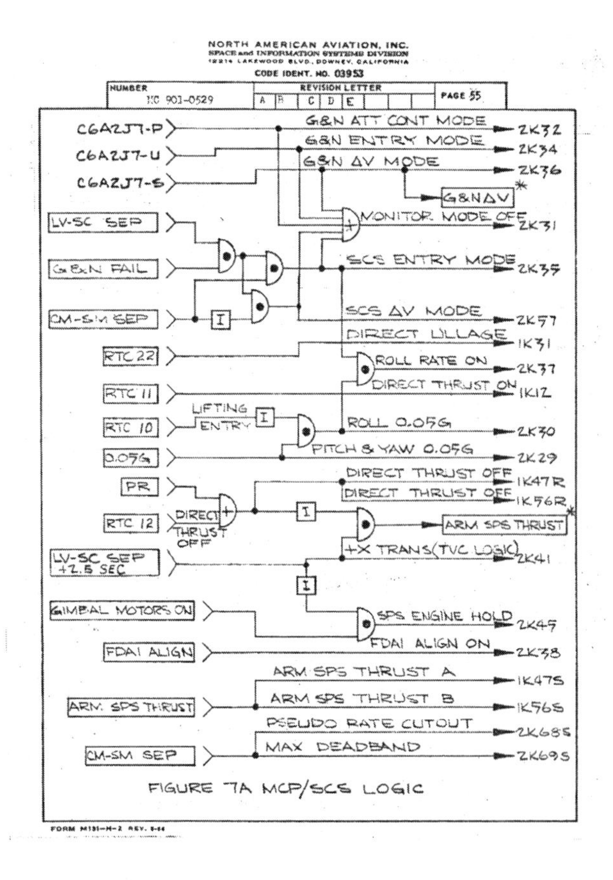

This is a digital logic page from the MCP Procurement Specification that shows the outputs to the Block I Stabilization and Control System and the complexity of the MCP (Courtesy of the Boeing Company).

Controller to become active. All the functions of the Ground Command Controller in the MCP were similar in design to the functions of the Ground Command Controller in AFRM 009. Like the AFRM 009 Control Programmer, the Mission Control Programmer controlled the attitude movements in space of the three airframes by commanding the Block I Stabilization and Control System to fire the Reaction Control System jets in the Service Module or to fire the Apollo capsule's Reaction Control System jets during Earth reentry. Whenever my Mission Control Programmer team had questions on the operation of the Stabilization and Control System, we would go talk to the major designers of the system: Paul Garcia, Ken Watson, Danny Moreno, or Mert Stiles.

At the time when my team took over the Mission Control Programmer, the design was already in progress. There was a definition of design released by Larry Hogan and Carl Conrad in an internal letter authored by John Beauregard, a Procurement Specification with a Specification Control Drawing released to the subcontractor, and subcontractor preliminary mechanization schematics—documents similar to those that described the AFRM 009 Control Programmer. However, there were two huge problems with the Mission Control Programmer that my team had to resolve. One was that no one on the program at Autonetics and at Downey was sure exactly what configuration of mechanization each Mission Control Programmer was at the time my team was given responsibility for its design; the second problem was that no one on the Mission Control Programmer program knew what configuration of mechanization the Mission Control Programmer should be on delivery to us in Downey. There were five Mission Control Programmers on order from Autonetics; three of them were of the production configuration for Spacecraft 011, 017 and 020, and two were to be submitted to the Qualification Test environments. In other words, the program was in shambles and out of control.

NORTH AMERICAN AVIATION, INC.
SPACE and INFORMATION SYSTEMS DIVISION
12214 LAKEWOOD BLVD., DOWNEY, CALIFORNIA
CODE IDENT. NO. 03953

NUMBER	REVISION LETTER								PAGE
MC 901-0529	A	B	C	D	E				15

3.3.4.5 MCP outputs. - The MCP shall provide output control functions as defined by Figures 15 and 16 for operating the following spacecraft subsystems during unmanned flight:

(a) Stabilization and Control System (SCS)
(b) Service Propulsion System (SPS)
(c) Telecommunications (T/C)
(d) Electrical Power System (EPS)
(e) Master Event Sequence Controller (MESC)
(f) Reaction Control System (RCS)
(g) Environmental Control System (ECS)
(h) Cryogenics System (CRYO)
(i) Up Righting System (URS)

3.3.4.6 MCP inputs. - The MCP will be provided inputs, to initiate output control functions, from the following:

(a) Guidance and Navigation (G&N)
(b) SIVB Instrumentation Unit (SIVB I.U.)
(c) Launch Control (LC)
(d) Up Data Link (UDL)
(e) Earth Landing System Sequence Controller (ELSC)
(f) Electrical Power System (EPS)

3.3.4.6.1 SIVB I.U. inputs. - The MCP shall accept input commands from the SIVB boost vehicle for sequence initiation as shown on Figures 5 and 16. The LV-SC Separation Start, LET Jettison, Lift Off, and SIVB Restart Command inputs will be derived from contacts located in the SIVB I.U. These contacts shall be armed with +28 vdc by the MCP and the return voltage will be utilized only by the MCP. All these command contacts will normally be open (source impedance greater than 1 megohm), and will close continuously (source impedance less than 10 ohms) upon command initiation for a minimum of 20 milliseconds. These contacts may remain closed until LV-SC separation is accomplished (approximately 2 seconds after LV-SC SEP "flag" origination) whereupon they will be opened. The MCP shall present an input impedance, with respect to MCP power return, of no less than 50 ohms for any SIVB I.U. input. Diode isolation shall be provided to all SIVB I.U. inputs in accordance with Figure 16D.

3.3.4.6.2 G&N inputs. - The MCP shall accept input commands from the G&N system for sequence initiation as shown on Figures 6 and 16. These input commands will be derived from relay contacts located in the G&N. The contacts shall be armed with +28 vdc by the MCP return voltage will be utilized only by the MCP. The command contacts will normally be open (source impedance greater than 1 megohm), and will be closed (source impedance less than 1 ohm) upon command initiation. The normally closed contacts of these G&N output relays may also be utilized as required by the MCP.

The G&N command inputs are shown drawn in the "OFF" condition on Figure 16. The MCP shall draw a maximum current of 0.5 amperes dc from any of these input commands, except as noted on Figure 16E.

FORM M131-H-2 REV. 9-64

This is another page from the MCP Procurement Specification showing the outputs to the MCP in paragraph 3.3.4.5 and the MCP inputs in paragraph 3.3.4.6 (Courtesy of the Boeing Company).

Resolving the two major Mission Control Programmer problems was no easy task, and accomplishing it required a tremendous amount of time and analysis from my team, Carl Conrad's team, the Autonetics team, and our management. For me and my team to be able to perform the tasks necessary, we had to study the details of the missions of each of the three spacecraft, study the Procurement Specification for the Mission Control Programmer and evaluate the subcontractor's Mission Control Programmer electronic schematics to comprehend the history of how the Mission Control Programmer was being mechanized and how it performed its functions. However, to get there, the first job at hand for us was to establish what the Mission Control Programmer design configuration should be to establish a design baseline. The next job was to evaluate the present configuration of each of the three Mission Control Programmers to make sure each one was of the proper configuration in agreement with the previously established design baseline.

The original release of the Mission Control Programmer Procurement Specification, which was given to Autonetics, established the baseline configuration of the Mission Control Programmer. All design changes made by us over and above the baseline MCP configuration defined by the original Procurement Specification that were required to be implemented by Autonetics were formally transmitted to them by a document called the Interdivisional Technical Agreement. My team spent much time reviewing the present configuration of each Mission Control Programmer and verifying the Interdivisional Technical Agreements that were incorporated in each of the five Mission Control Programmers. Then the team reviewed the released Interdivisional Technical Agreements for implementation to understand each one and then determine which Interdivisional Technical Agreement had already been incorporated into each Mission Control Programmer. It took my team, in cooperation with Carl Conrad's team and the Autonetics team, months of work and many meetings to complete the separate documentations. Then the team was back in business, and from here on out the process entailed numerous meetings with the subcontractor team and with AFRM 011, 017 and 020 responsible engineers and management in Downey. These meetings almost always resulted in action items being assigned to several of the attendees. The answers to these action items on their due dates were in the form of an internal letter, requiring a telephone call or a verbal answer during the next meeting to the person who needed the information resulting from the action item.

There were many internal letters released, meetings attended, and telephone calls made over the course of 1965 and 1966. Several examples will be given in the narrative that follows. Because the PC had not been invented yet, the main methods of communication were the written letter and the telephone. Meetings were working meetings, engineering problem solution meetings, and management program status or program review meetings. As in the case of the Control Programmer for AFRM 009, the Mission Control Programmer was being manufactured by Autonetics in Anaheim, and my team again had to spend quite a bit of time at Autonetics to ensure that the Mission Control Programmer met all design requirements and that the delivered systems could perform the jobs they were designed to do. Instead of dealing with John Rowe for mechanization of the Mission Control Programmer, my team now had to communicate with Junior Jacobs (the lead Autonetics designer for the Mission Control Programmer) and with Cal Smith (project engineer for the Mission Control Programmer) instead of Jerry Paccassi. The Autonetics program manager for the Mission Control

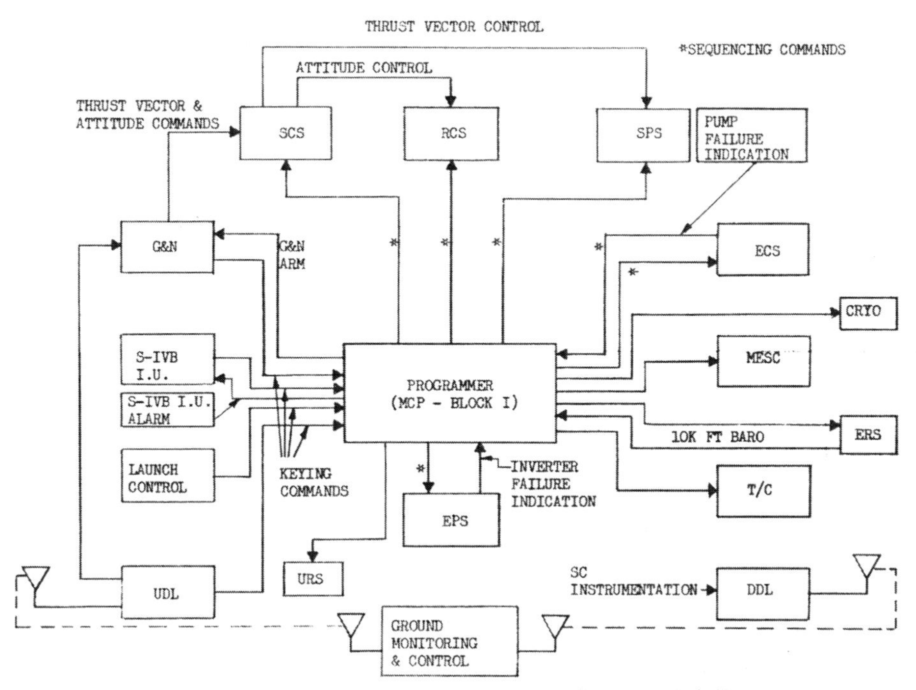

Figure 1. Mission Control Programmer Block Diagram

This drawing depicts MCP interfaces with other Apollo systems during the missions of AFRMs 011, 017 and 020 (Courtesy of the Boeing Company).

Programmer was Don McLean, who had also managed the Control Programmer (a plus for me and my team).

During the first Autonetics and Fouts' MCP team management review meeting that I attended, Bill Fouts announced to all the attendees that I was taking over the Mission Control Programmer Design Project. A discussion followed on what was happening in the program and where the program was relative to the schedule. Before the discussion got really going, and because there were Autonetics personnel present with whom I had never dealt before (such as Cal Smith, the project engineer), I asked, "How are we going to operate between us and Autonetics during the Mission Control Programmer program?" Dick Ellstrom, who was the Mission Control Programmer designated subsystem project manager at S&ID, was the first to answer: "What's wrong with the way you've been operating on the Control Programmer? Are you all of a sudden turning chicken?" Dick looked around the room and asked, "Does anyone have an objection to that, that we operate the same as in the Control Programmer program?" Everyone agreed to continue operating as we had done for the AFRM 009 Control Programmer, and I indicated that it was fine with me to continue that way because I did not want to lose my ability to personally give Autonetics on-the-spot engineering written direction. That was exactly the answer I was waiting to hear because I felt that if I could personally give Autonetics written engineering direction, as was done on the Control Programmer, and if I could personally team up with Junior Jacobs on any MCP redesign as with John Rowe, we had a chance of completing the scheduled Mission Control Programmer milestones.

Because the missions of the next three unmanned Apollo spacecraft were different, the Mission Control Programmer design team had to make sure that the Mission Control Programmer was mechanized and manufactured to fulfill each of the three spacecraft's missions. My team spent days, evenings and sometimes weekends at Autonetics meeting with Junior Jacobs to review the design and with the Autonetics test teams to make sure that the Mission Control Programmer was built and tested correctly and to give Autonetics engineers on-the-spot engineering direction and approval of needed design and test changes.

To help readers understand the problems encountered in the design and mechanization of the Mission Control Programmer, here is my attempt to put the different missions in laymen's terms as best I can: In the Apollo Block II Command Modules that would be qualified for manned moon flights, the operation of the Guidance Navigation and Control System was mandatory for a successful moon mission. This system was known as the Primary Guidance Navigation and Control System, and it was fully automatic, a development made possible by a digital computer designed by Eagle Knight Dr. Ramon Alonso, which was called the Command Module Computer (or CMC) and contained the software application programs (now called apps) to navigate the Apollo capsule to the moon and return it safely back to Earth. It was called primary because there was also a secondary or backup system, the Block II Stabilization and Control System, that could replace the Primary Guidance Navigation and Control System if the primary system failed to operate properly. The Block II Stabilization and Control System was not an automatic system; it was designed to be operated manually by the astronauts. The Stabilization and Control System was the system that was used in AFRM 009 to control the spacecraft. However, the AFRM 009 Stabilization and Control System was a Block I system, not Block II. The Block I system was used because it was capable of controlling the spacecraft in a suborbital shot or even manned Earth orbital mission, but it did not have the required reliability for a moon mission. The Block I Stabilization and Control System was primary for the Block I airframes, and the Block II system was at this time being designed for the Block II Apollo capsules as a backup system. The missions for the last three unmanned spacecraft were more complex than the AFRM 009 mission, and even though they were also Block I vehicles and did not contain a Primary Guidance Navigation and Control System, they did have a Stabilization and Control System that was primary that allowed the Apollo guidance, navigation, and control techniques to be verified. (In later chapters, I will discuss in detail the reasons why the Block I Stabilization and Control System was not suitable for moon missions.)

12

Final MCP Design for AFRMs 011, 017 and 020

Even though Spacecraft 011 was going to be a suborbital mission, unlike AFRM 009 (which was recovered from the Atlantic Ocean off the coast of Africa), Spacecraft 011 would be recovered from the Pacific Ocean about 800 kilometers away from Wake Island. This mission would test the restart capability of the Service Propulsion System engine in the Service Module. This restart capability was a very critical part of a lunar mission because the engine had to fire more than once so that it could make course corrections on the way to the moon and on the way back to Earth. The engine also had to fire as the Apollo spacecraft approached the moon to place the spacecraft into lunar elliptical orbit, fire again to circularize the orbit and then again whenever the Apollo crew was ready to blast away from moon orbit and head for home. For any of these maneuvers to be made correctly, the Service Propulsion System engine had to be pointing through the center of gravity of the combined Apollo Command Module–Service Module spacecraft (the Service Module carried the tanks with oxygen, fuel for rocket engines, nitrogen for pressurizing fuel tanks, and other consumables, and as these consumables were used, the center of gravity shifted, requiring the nozzle to be repositioned every time the engine was fired). If the engine was not pointing through the center of gravity when it fired, the Command Module–Service Module combination would tumble instead of accelerating or decelerating as required, which would cause the spacecraft to go out of control. Another concept that is important to Apollo, or to any spacecraft entering Earth's atmosphere from outer space, is rolling entry (described in detail in a previous chapter). This maneuver is what the Apollo spacecraft was commanded to perform during Earth reentry to determine the Apollo capsule's landing point in the ocean, a task accomplished by dissipating spacecraft energy.

Spacecraft 017 was going to perform an Earth orbital mission, which was more complex than the Spacecraft 011 suborbital ballistic trajectory mission. The Spacecraft 011 (and, for that matter, the AFRM 009) trajectory was called a ballistic trajectory because it was like shooting a rifle pointing up at about thirty degrees in the air, in which case the bullet will follow a ballistic trajectory as it falls to the ground. For the Spacecraft 017 mission, the guidance navigation and control techniques were necessary and more complex and required more accuracy to ensure mission success because the spacecraft was being guided to a specific point in space so it could achieve a specific altitude to orbit around Earth. The Block I AFRM 017 Stabilization and Control System had the ability to perform all those functions in an Earth orbital mission. In the Block II moon-mission-capable spacecraft, these functions would be performed by the

Primary Guidance Navigation and Control System. One of the critical elements of this system was the Inertial Measurement Unit, which was very accurate and made the Primary Guidance Navigation and Control System capable of navigating the Apollo spacecraft to the moon and returning to Earth. By contrast, the Block I Stabilization and Control System in AFRM 017 did not have an Inertial Measurement Unit. Instead, it had six body-mounted attitude gyros (BMAGs, pronounced "B-Mags"), which performed the same functions that the Inertial Measurement Unit did in Block II capsules. The BMAGs were needed in ARFMs 017 and 020 for navigation during their missions. However, the BMAGs would soon drift out of alignment because they were not of the same caliber as the Inertial Measurement Unit; for this reason, the BMAGs had to be periodically realigned through electronic means so that they could maintain their accuracy.

Assuming the earlier flights were successful, the mission of Spacecraft 020 would be to leave Earth's orbit and go into deep space toward the moon and then return to Earth in a rolling entry maneuver. AFRM 020 would also certify that the guidance and navigation techniques could hold the Apollo capsule inside the entry corridor and enter Earth at the exact entry point. The spacecraft had to be fully equipped with all systems ready as if it were going to the moon because the next spacecraft to go into space after Spacecraft 020 would be a manned Earth orbital mission, which would be followed by a moonshot in which the spacecraft would swing around the moon and return to Earth. A spacecraft mission with three astronauts going around the moon was traumatic enough because it had never been done before. So, let us get back to the internal letters, meetings and telephone calls that were necessary to get the Mission Control Programmer problems solved.

There were numerous meetings, internal letters, and telephone calls necessary for the Mission Control Programmer team to solve the two main problems with the Mission Control Programmer; the following are some typical examples. Several internal letters were issued by Larry Hogan's unit to Bill Paxton's unit regarding the requirements for alignment of the BMAGs. As mentioned earlier, the Block I capsules had no Inertial Measurement Unit for navigation like the Block II capsules but instead had a set of BMAGs that tended to drift and had to be realigned periodically. There were several internal letters issued by Larry Hogan's unit describing the requirements for alignment of the BMAGs in Spacecraft 011, 017 and 020. One internal letter issued on February 10, 1965, requested support from the other Apollo units by forwarding information to Larry's unit so his unit could perform an analysis to determine how often the BMAGs should be realigned. A second internal letter issued on March 11, 1965, also by Larry's unit, set down the policy for aligning the BMAGs in AFRMs 011, 017 and 020. The alignment requirements were to align one minute before lift-off, immediately after boost, and periodically once an hour during the coast phase for Spacecraft 017 and 020 only. Another internal letter (one of many that were written by me to Don McLean) was written on January 16, 1966, while I was visiting Autonetics, giving Autonetics engineering direction for setting the values of the gimbal positions for the various Service Propulsion System engine firings during the Mission Control Programmer mission.

On June 13, 1966, Bill Paxton issued an internal letter leaving me as acting supervisor while he went on a vacation for two weeks. (More will be written about this letter later.) The meetings held, as documented in my notes, were of thirteen types: meetings with Autonetics in general, meetings called by Cal Smith in his office, meetings called by Don McLean in his office, meetings called by Bob Antletz in his office, meetings called by Vaughn Slaughter in his office, meetings called by Bill Fouts in his office, meetings called by Fred Schmoldt in his office, meetings called by Norm Rassmussen in his office, S&ID general meetings, program review meetings, Engineering Review Board meetings, NASA meetings, and cost review meetings.

The first meeting with Autonetics on record was one on January 4, 1966, to discuss several major Mission Control Programmer problems that required solutions as soon as possible. Those in attendance from our company were Gene Rucker, Don Farmer, Vince Merick, Carl Conrad, and me; Autonetics was represented by Junior Jacobs and Walter Farley. There was a problem with relays not having well-defined characteristics from the supplier. Autonetics was working on this problem and was going to take care of it. Another problem concerned the electronic circuit in the Ground Command Controller of the Mission Control Programmer that received the ground real-time commands through the up-data link in the Mission Control Programmer backup mode. At the time of this meeting, no one really knew how big the problem was because there was conflicting information regarding this issue. We called it the UDL problem because it involved the up-data link from the ground equipment, and because of the unresolved status of this problem, several action items were assigned to Autonetics.

On Tuesday, January 18, 1966, I spent the day at Autonetics. I was there to consult with Junior Jacobs and to oversee the mechanization of changes to the Mission Control Programmer. Early in the day, I had a meeting with Project Engineer Cal Smith in his office to discuss the UDL problem in detail. By this time, Autonetics had been able to completely characterize the UDL problem, and Cal was presenting some proposed "fixes" to me. Cal said that the preferred fix had a high risk because it entailed adding a new component on an electronic board of the Mission Control Programmer that might or might not be able to accommodate this change. Autonetics was going to study the addition of the new electronic component, and then they would write an Interdivisional Technical Agreement that proposed a solution to the problem. All Mission Control Programmer design changes were documented on Interdivisional Technical Agreements, as mentioned before, and were identified by the IDWA number followed by four numbers or by a request for change. If the Interdivisional Technical Agreement was revised, then it received a letter (such as A, B and so on) depending on the number of times it was revised. A typical Interdivisional Technical Agreement number would be Interdivisional Technical Agreement 6533-0002. In discussing or referring to each Interdivisional Technical Agreement, it was standard procedure for us to use only the last four digits since the IDWA number was always the same for the Mission Control Programmer. Thus, Interdivisional Technical Agreement 6533-0002 became Interdivisional Technical Agreement 0002. Other problems I discussed with Cal Smith in our meeting were two important Interdivisional Technical Agreements that would result in a design change to the Mission Control Programmer and various other issues that required solutions.

Later that day, I and others met with Don McLean in his office to discuss the impact that the UDL fix would have on the scheduled delivery of the Mission Control Programmers. Don indicated that the UDL fix was documented in an Interdivisional Technical Agreement and that Autonetics needed go-ahead from us to implement it. However, Autonetics would have to know what the alternate plans would be in case adding the new component on the electronic board did not work, and they would also have to assess the Mission Control Programmer delivery schedule impact before they really needed the go-ahead from us. Early that evening, I called Bill Paxton and informed him of the UDL fix Interdivisional Technical Agreement 0021. I told Bill that I had reviewed the technical agreement, and everything looked okay. Bill then informed me that a team was being formed in Downey to spend a week on design review of the Mission Control Programmer. The team captain was going to be Carl Conrad. Team members would be Gene Rucker, Don Farmer, and Mike Falco. Bill said that a conference room in Building 318 had been reserved for the effort. He also told me that Vince Merick and I were to ramrod the UDL fix (and all testing associated with it) in Downey. In addition, he said I was to attend a meeting the following day at 9:00 a.m. in Don McLean's office to discuss the Mission Control Programmer factory test equipment and its impact on Mission Control Programmer qualification testing. As soon as I got off the phone with Bill, I followed it up with a call to Gene Rucker to inform him that he was to be involved in the Mission Control Programmer design review and then a phone call to Don Farmer to tell him that he also was to be a member of the design review team.

At around 8:30 p.m. that evening I was informed by Autonetics personnel that Mission Control Programmer Production Number 1 was being presented to the Air Force Quality Control for acceptance and would probably be shipped to Downey that same day, so I gave Autonetics instructions on the exact address to which the Mission Control Programmer was to be shipped. By that time, I was tired and went home, knowing that I would be returning to Autonetics the following day. This was a typical day for me during the Mission Control Programmer program.

The following day, I attended Don McLean's 9:00 a.m. meeting at Autonetics. Any time Don held a meeting, Autonetics management was well represented. Autonetics personnel who were normally present were Norm Rassmussen, who was head of manufacturing; Bob Fettes, supervisor of Mission Control Programmer design; Fred Schmoldt, project engineer for Mission Control Programmer testing; Cal Smith, Mission Control Programmer project engineer; and Vaughn Slaughter from Mission Control Programmer project engineering. The only two who were not part of management were Vaughn Slaughter and me. The meeting on this day was a long one because of the many problems facing the Mission Control Programmer. It started with a briefing by Norm Rassmussen on the delivery schedule for the Mission Control Programmer. Delivery schedules were always topics that interested us deeply because our teams in Downey had many meetings with Bill Fouts on this same subject, and my team was mindful of the importance of delivering the Mission Control Programmers in time to support our delivery date of the spacecraft to NASA. Attendees at Don's meeting discussed the manufacturing test equipment and its ability to support the testing of the Mission Control Programmers. Mission Control Programmer Life Test and Qualification Tests were another topic that came up because there were problems that had to be resolved in that part of the program. Then followed a lengthy discussion on the plan for resolving the UDL problem and the need for some decisions regarding the fix to some of the Mission Control Programmers. At the end of the meeting, I picked up two action items: one was to write a formal letter to be channeled through Bill Paxton and the IDWA group on a test waiver requested by Autonetics, and the second was to write an Interdivisional Technical Agreement for a design change to the Mission Control Programmer. The two action items that had been assigned to me would suffer the normal Bill Paxton delay, causing my already-full schedule for the next few days to overflow coupled with the agony of their impending delay when channeled through Bill.

By March we were making headway on determining the current mechanization configuration of each Mission Control Programmer, as well as the mechanization configuration that each Mission Control Programmer should have when delivered to us. A meeting was held at Autonetics on March 4 to discuss all the Interdivisional Technical Agreements that defined mandatory design changes to the Mission Control Programmer. Five days later, a meeting was convened at our facility in Downey in Building 318 to discuss and determine the mandatory Interdivisional Technical Agreement design changes to the Mission Control Programmer for Spacecraft 011. Cal Smith then called a meeting in his office the following day to finalize the Interdivisional Technical Agreement design changes that had to be implemented for each Mission Control Programmer prior to delivery to us in Downey. There were many more changes that were to be made to the Mission Control Programmer, and more problems that had to be solved, but in the end Autonetics delivered Mission Control Programmer systems on schedule that met the needs of AFRMs 011, 017 and 020.

Due to the distinct missions of each spacecraft, each MCP was slightly different in its design. The difference for the MCPs was defined by implementing different Interdivisional Technical Agreements into each of the three production MCPs. To identify each MCP, a different dash number was added behind the basic part number. For example, all three MCPs had the same basic part number, but AFRM 011 had -001 added behind the basic part number, AFRM 017 was identified by -002, and AFRM 020 was -003.

Around the second half of May 1966, it seemed that the major design problems for the Mission Control Programmer had been solved and the pressure of the job was decreasing. I began to feel restless and sort of relieved and maybe a little as if I were on a boat out at sea without a sail or a rudder—kind of aimlessly floating. I always felt like the pressure of the work on the Control Programmer and the Mission Control Programmer was having no effect on me, but boy, was I wrong. Weeks earlier, my stomach had started to bother me; I could often feel it gurgling and churning away. I made an appointment to see a doctor located on Brookshire Avenue in Downey, just a couple of blocks behind Building 318. The doctor put me through the whole upper and lower GI procedure, and he finally told me that he could find nothing physically wrong with me. The doctor told me that the symptoms could stem from job-related pressure or an ongoing divorce. Because I was not going through a divorce, we talked about my job. The doctor said, "There are two solutions to your problem: one is to remove the pressure from your job, and number two is to remove yourself from the pressure." My problem became so unbearable that the only solution was to remove myself from the pressure. I decided that what I needed was a change in jobs with new people, new management and new projects, so I mailed out my resume to the different companies in the Los Angeles area that had placed ads for engineers in the employment section of the *Los Angeles Times*.

Early in the month of June, I received a phone call from an engineering manager at Aerojet General Corporation in Azusa. The manager told me he had received my resume from his Human Resources Department and wanted to know whether it was true that I had worked on the Apollo Spacecraft 009. I told him that yes, it was true, and he followed up by asking whether I was interested in going to work for Aerojet. The job description given over the phone was for a technical specialist on a new project that was described in detail, and the job could eventually evolve into a management position. It all sounded good, but I was not quite sure I was ready to leave my current position. I told the Aerojet manager to give me some time to think about the offer and a chance to talk to my wife about it. I said I would get back to him within a week.

Sometime before Bill Paxton went on vacation, he came to me and told me that one of the members on my team was going to be laid off, so he point-blank asked me which one of the two I would like to hold on to: Gene or Don. I indicated to Bill that I preferred to hold on to Gene. This question surprised me because ordinarily a supervisor would never ask his lead engineers to decide who was going to be laid off, as doing so would immediately divulge that a layoff was imminent and rumors would start to fly, so I kept this news under my hat for fear that it would start a panic. The method used by most supervisors and managers to lay off employees was to request a list from the lead engineers that would rank employees under their leadership, starting at number 1 for the highest-ranked engineer, followed by number 2 for the next-ranked engineer, and so forth. The list would be updated regularly, and whenever a layoff was declared, the supervisor or manager would use these lists to select the layoff candidates without consulting anyone. This episode of Bill Paxton asking me straight out which one person should be laid off did not set well with me and was the time my stomach problems started.

I knew that things were slowing down, and engineers were being laid off in almost every design unit. Don Farmer was eventually told by Bill Paxton, sometime in early June 1966, that he was being fired. Fortunately for Don, he was given enough time to find a position at the Columbus Division in Ohio. Don was lucky there because the person who hired him, in reviewing Don's resume, mistook Chico State (Don's alma mater) for Ohio State and on that basis granted Don an interview. Once the interview was completed, the manager was so impressed with Don that it did not much matter whether he had graduated from Chico State or Ohio State. Don had the job, and he and his wife moved to Columbus, Ohio. Don was a pilot who kept his aircraft parked at the Long Beach airport. He and his wife just packed their belongings in the plane and flew to Ohio.

Don's airplane was a single-engine plane, and one day, several months before he departed for Columbus, he asked Gene Rucker and me whether we would like to go for a ride in his plane. Gene and I both said we would love to go for a short flight. The three of us left work, got into Don's car and headed down Lakewood Boulevard for the Long Beach Airport, which was not far from Downey. While we were about three or four thousand feet up in the air, flying over the Pacific Ocean, we could see Long Beach, the San Pedro Harbor and out toward Los Angeles. From where we were, it was such a great view, and then the thought struck me: "What if there is an air mishap and the plane crashes in the ocean? The three of us are supposed to be at work, and I am responsible for these two guys. If we die, will our company insurance pay off, or will the company litigate the whole issue?" Right there and then, I could not wait to get down on the ground.

13

Farewell to the Mission Control Programmer

On June 13, Bill Paxton went on vacation and left me as acting supervisor from June 13 to July 4, 1966, as was mentioned in an earlier chapter. On June 15, Gene Holloway, the NASA Mission Control programmer subsystem manager, was in town for a meeting in which he and I discussed a failure of the electrical power system in Spacecraft 011 at the Kennedy Space Center that did not involve the Mission Control Programmer, along with the status and number of spare automated systems for Spacecraft 017 and 020. Also discussed were the exact voltage settings for the Mission Control Programmer gimbal position for each of the Service Propulsion System Engine firings of the remaining two unmanned spacecraft. The gimbal position voltage was the setting that determined the position of the Service Module Propulsion Engine nozzle for the correct firing of the engine. The nozzle was driven to that position and held there by a complicated magnetic clutch and braking servo system that was the area of expertise of Eagle Knight George Cortes. The following day, there was a scheduled Mission Control Programmer cost review meeting at Autonetics. The cost review meeting attended by Gene Holloway and me at Autonetics on June 16 was a detailed review of the Mission Control Programmer expenditures by month, the number and level of technical individuals involved in the program by month in each activity in the Mission Control Programmer design, development and manufacturing activities and their cost. The price of all materials used and where they were used was also detailed. The meeting lasted most of the day, and Gene went home to Houston with plenty of information on how much the Mission Control Programmer program would likely cost NASA at the end of the program completion and where the money was being spent.

It was during this period, after Gene Holloway left town and before Bill Paxton returned from vacation, that I was offered a promotion by my Hound Dog cruise missile mentor and supervisor Ed Kelley in the Apollo Stabilization and Control System organization. Ed was the Stabilization and Control System assistant designated subsystem project manager, working directly for Bill Fouts, who was now the Guidance and Control Department manager. In addition, Bill Fouts held the title of Stabilization and Control System designated subsystem project manager, responsible for managing the day-to-day activities of Honeywell in Minneapolis, Minnesota. Honeywell was mechanizing, manufacturing, and testing the Block II Stabilization and Control System for us, similar to how Autonetics had mechanized the Control Programmer and the Mission Control Programmer for S&ID. The Block II Stabilization and Control System was to be the backup system in the Block II Apollo Command Modules. Bill had a full plate on his hands with the duties of department manager and no time to devote to the daily activities of the Block II Stabilization and Control System procurement from Honeywell. Bill's solution was to promote Ed Kelley to be his assistant designated subsystem project manager, and he gave Ed full authority and responsibility of the designated subsystem project manager office.

This is how the promotion offer was made to me: I was called into Ed's office and asked whether I was interested in being the Block II Stabilization and Control System resident representative at Honeywell in Minneapolis. Ed told me it was a promotion but required a move to Minneapolis, Minnesota, in July 1966. He also informed me that the two engineering resident representatives at Honeywell at that time were Wes Jensen and Bill De Viney, but Bill was coming back to Downey, which would leave one open position. The assignment would be for around nine months—certainly no longer than one year. Ed said he was not sure that it would be offered to me, but he wanted to know whether I was interested in case the offer was made. I told Ed that I was not sure because since it involved a move out of state; I would have to clear it with my wife, so I told Ed that I would defer a yes or no answer until I talked to her. I promised to give him an answer within the next two days. The following day, I let Ed know that I was interested in the job.

Several days later, Ed called me into his office. We sat down and chatted for a bit, and then he told me that the job was mine if I still wanted it. Ed said he and Bob Epple, the department assistant manager, had been trying to decide between Bill Cavanaugh (who worked in the Stabilization and Control System Design unit) and me for the

The two large boxes are part of the Mission Control Programmer being tested in the lab for AFRMs 011, 017 and 020 (Courtesy of the Boeing Company).

job. They discussed, evaluated, and struggled for several days, and finally one day Ed told Bob, "Bob, you and I know full well who it should be for the job!" Bob's reply was "Yeah, I know." Ed told him, "Do you want to tell him or should I?" Bob's response: "You tell him." I told Ed that I would take the position on the condition that it would be no longer than one year because that was the only condition my wife had when she agreed to move to Minneapolis. Ed assured me it was guaranteed to be no more than one year. He said the job was mine and that the transfer paperwork would be processed immediately. That same day, I called the manager at Aerojet General and informed him that I would not be accepting his job offer. Days after I had accepted the new promotion, my stomach problems disappeared.

A view of the MCADS for AFRMs 011, 017 and 020 (Courtesy of the Boeing Company).

As the engineering resident representative at Honeywell, I would be working on the Apollo Block II Stabilization and Control System for the manned Apollo capsules designed to take astronauts to the moon and return them safely home. For me, it was a radical change to be working on the manned capsules, but that is where I wanted to be. In this position, I was to be the "Johnny-on-the-spot" responsible for reviewing and approving (or disapproving) any Stabilization and Control System design changes proposed by Honeywell; it was remarkably like the job I was leaving behind on the Mission Control Programmer except that this assignment was on the manned Apollo capsules and I was not responsible for the design of the system. By July 1966, most of the Stabilization and Control System for all Block I capsules had already been delivered to S&ID in Downey. Even though all Block I capsules had not yet flown, the design of the Block II Stabilization and Control System for the future Block II capsules had already begun in order to meet the moon landing mission by the end of the decade. It was my task to ensure that the Block II Stabilization and Control System items met all specification requirements by intimately associating myself with the Honeywell management and designers. Because there was still some work left to be done on the Block I Stabilization and Control System, as the engineering resident representative, I was also responsible for overseeing the completion of that effort.

The Block II Apollo capsules were the Command Modules that were going to make lunar flights manned by three astronauts, starting with Apollo 7 and on up the line to Apollo 17. The Block II Stabilization and Control Systems for the Block II Command Modules were in the final design stage and qualification testing at Honeywell. There was still much work to be done before the Block II Stabilization and Control System was ready for an Earth orbit or a moon flight. The Apollo 012 Command Module Block I Stabilization and Control System had already been delivered to NASA and installed in the capsule. (Apollo 012, if you recall, was a Block I Command Module scheduled to be the first manned Apollo flight and would fly an Earth orbit mission with the Block I Stabilization and Control System as the Primary Guidance, Navigation and Control System.)

When Bill Paxton returned from vacation, I informed him of my promotion. Bill was exceedingly upset, and I could see fear in his eyes because I was leaving his unit, but there was not much he could do about it. To relieve his fear, I told Bill that he at least had Thom Brown, who was more than capable of replacing me as lead engineer on the Mission Control Programmer because the design of the MCP was now completed; from that point, the tasks remaining were to ensure that the MCPs delivered to us by Autonetics met the Procurement Specification and that the MCP worked properly in each of the three unmanned spacecraft.

On Saturday, July 23, 1966, I flew out to Minneapolis on a twelve-day trip to meet the North American Aviation cadre working at Honeywell on the Stabilization and Control System. I would also meet the Honeywell management and engineers on the project. While I was there, I could scout the area for a place to live when my family and I moved there, and I could size up the job to decide for sure whether it was really for me. To help me make up my mind, I was carrying some letters that had been written by several S&ID personnel who had served as resident representatives in Minneapolis for North American S&ID. I intended to read these letters while I was in Minneapolis because they described the additional costs of housing due to the need to rent or lease a place to live, as well as the added expense of heating a home during the long winters. There were further costs brought on by winter conditions unknown to representatives from California, such as winterizing the car, adding snow tires, and possibly replacing the old battery with a new battery, to say nothing of the new winter clothing needed by the whole family to withstand the bitter cold. Because of the temporary assignment, there were also Minnesota non-resident state income taxes to pay and many additional hours of overtime that had to be worked without pay. The argument of the letters was an attempt to justify why the company should provide representatives with additional pay of $115 a month while working in Minneapolis.

When I read the letters about the extra $115 a month, the information and arguments presented did not really click in my head because, first, I felt that maybe it was all being exaggerated and, second, I had never lived through a snowy and cold winter and really had no idea how bad it could get. My visit to Minneapolis was in the summertime, which was not so bad, and so a cold climate meant nothing to me. It was not until we had moved to Minneapolis and had to survive blizzards, freezing rain and eight months of winter with below-zero temperatures that I reread the letters, and it all made sense to me then, but by that time it was too late to renege on the agreed-to assignment.

Before my departure, on the first visit to Minneapolis, I was told that if I did not like the job setup, I could renege on the decision to take the promotion. However, the city was beautiful, and the job seemed to present a challenge, so I was not about to turn the offer down. I was later to learn, after moving there, that being a visitor to Minneapolis in summer or winter can in no stretch of the imagination be compared to living there.

When I arrived at Honeywell on Monday morning, July 25, 1966, I was given a temporary badge and then taken to the Rockwell offices, where I met Sterling Shelby, the materiel representative who was in charge of the S&ID Office; Jeannie, who was the secretary and worked for Honeywell; Wes Jensen, the S&ID engineering representative; and Ted Nakashima and Marv Smith, the two quality control representatives for S&ID. I was then taken around the plant by Wes Jensen, who introduced me to Al Hammel, the Block I SCS System section head; John Bancroft, the Block I System project engineer; and Leon G. Christensen, Project Office project engineer. Most of the morning was get-acquainted time, so Wes and I just went around meeting some of the design engineers and production engineers. Later in the day, Al Hammel had some Block I Stabilization and Control System problems he wanted to discuss with the Stabilization and Control System Design Unit in Downey, so he invited Wes and me for a conference telephone call to Downey. On the Downey side for the discussion was Carl Conrad, the unit supervisor, accompanied by Bill Cavanaugh. One of the problems under discussion concerned AFRM 011. It had been determined that the Service Propulsion System engine clutch had high electrical current that made one of the Stabilization and Control System electronic boxes suspect with a bad power supply. The second problem concerned the Stabilization and Control System in Spacecraft 012 (Apollo 1), which was to be the first manned Apollo mission. My notes from that date are incomplete, but in all cases the problems were of a failed component nature and not engineering design technical problems.

Tuesday, July 26, was very much like the day before, with familiarity discussions being held in the S&ID offices. At 10:00 a.m., Wes Jensen, Sterling Shelby, Dick McConnel (the S&ID administrative buyer), and I attended a Block I program review meeting with Leon Christensen, Leslie Bankson, Paul Begler, Irv Fong, and Max Thompson, all of Honeywell. All the major problems facing the Block I Stabilization and Control System were discussed at length. At 2:00 p.m., Wally Lundahl, the chief engineer, held a pre–Dale Myers meeting in his office. Dale, the Rockwell VP Apollo program manager, was coming in the following day for a briefing on the status of the Block I and Block II Stabilization and Control System Programs. At 3:20 p.m., Wes and I met with Roger Britt and Walt Melin of Honeywell to discuss the Block I sustaining engineering for the Block I Stabilization and Control System. The need for the sustaining engineering had been agreed by Honeywell and S&ID, but Honeywell claimed that the two designers currently assigned to the system were not enough to handle the problems. We decided to bump this problem up to the meeting with Dale Myers for a decision on the following day.

On Wednesday, July 27, Ed Kelley, and Dick McConnel picked up Dale Myers at the Minneapolis airport and drove him to the Honeywell plant. The Apollo program status meeting was held to discuss the most critical problems that Honeywell was facing in the Block II program (as well as some necessary cleanup work on the Block I Stabilization and Control System) and to introduce me to the Honeywell management. Present at this meeting were Corliss Perkins, the Honeywell VP in charge of the Block II program; several Honeywell division directors; the Honeywell chief engineer; Apollo Honeywell engineering managers; Honeywell project engineers; Dale Myers; Jim Edwards, who was the S&ID director of materiel; and Ed Kelley, my boss, who was the Stabilization and Control System assistant designated subsystem project manager. The meeting lasted from 8:30 a.m. to 4:30 p.m. I was introduced at the beginning of the meeting, followed by a discussion on what my authority would be. There was concern by some Honeywell managers because they felt that the engineer I was replacing, Wes Jensen, did not have authority in certain areas that were now crucial for the success of the program. The Honeywell high-level managers made it clear that if I was not to have authority in these areas, they wanted an engineer who did. Ed Kelley replied that he had already spoken to the Honeywell Stabilization and Control System chief engineer and had told him that I had authority in anything that involved North American Aviation, without needing to first consult the home office in Downey, and anything I signed was binding on the company. Those words certainly clarified what my position and authority would be at Honeywell.

Unlike most meetings that last all day, instead of going out to lunch at a restaurant, we were served lunch in the Honeywell executive dining room, which at the time was a treat in corporate America. That afternoon, after the meeting ended, I returned to my hotel for a couple of hours and then tried to get hold of Ed to have some dinner. I could not find him, so I went out and had dinner alone. The following day, Ed told me he had gone to dinner with Dale Myers and had been looking for me but could not locate me, so I had missed an opportunity to break bread with Dale.

On Thursday, July 28, I attended the monthly coordination meeting, where I met Chuck Feledy, the Honeywell senior administrative buyer. The rest of the week and part of the following week, I was immersed in attending meetings and discussing Block II Stabilization and Control System problems with the Honeywell personnel. Many times during the days that I was there, there were conference telephone calls between Honeywell and the Downey engineers in which I was invited to participate. Mostly, though, I used this time to learn the ropes around Honeywell and make more personal contact with the Block II Stabilization and Control System design engineers with whom I was going to work closely during this assignment. I also spent much time studying the operation of the Block II Stabilization and Control System because, despite my familiarity with the Block I Stabilization and Control System, there were many differences between them; whereas Block I was a primary system, the Block II version was a backup system to the Block II Primary Guidance, Navigation and Control System. On August 4, 1966, I boarded a plane and headed for home.

14

The Honeywell Resident Representative Assignment

On returning to California on August 4, 1966, and reporting to work in Downey the following day, I felt as though the trip had been a great success, so I decided to take the promotion and move to Minneapolis. The question for our family was what to do about the house we were buying in Santa Fe Springs, California, when we moved to Minneapolis. The company was going to pay for moving the furniture to Minneapolis, and we certainly did not want to leave the house vacant for one whole year. One option was to rent it, but that was not a good choice with our family being so far away in Minnesota. However, if we sold the house, the company would reimburse us the realty fee, so in August, after I returned from the Honeywell trip, we put the house up for sale.

Up until the day before we left for Minneapolis, my priority was getting up to speed on the Block II Stabilization and Control System. So exactly what was the function of the Block II Stabilization and Control System in the Apollo program? The unmanned Apollo capsules, as well as AFRM 012 (the first manned Apollo flight, which was named Apollo 1), were all Block I capsules. In the Block I/Block II Apollo capsule concept, the Block I designation meant that this group of capsules would be identical to the original capsule design and good only for Earth orbit. The Block I capsules used the Block I Stabilization and Control System as the primary means of guiding, navigating, and controlling the maneuvers of the Apollo capsule in space and during Earth reentry. In contrast, the Block II Apollo capsules had moon mission capability and used a different method of controlling capsule maneuvers. The design of the inner and outer structures of the Block I and Block II Apollo capsules were identical, but they differed in the design of some of the electronic systems.

Normally in the Block II capsules, control of a manned Apollo capsule would be accomplished by the Primary Guidance, Navigation and Control System, which was fully automatic thanks to Dr. Ramon Alonso's digital computer, which orchestrated the whole Apollo flight with minimal astronaut intervention from the time the capsule was launched into space to its final splashdown in the ocean at the completion of the mission. (One of Ramon Alonso's most brilliant innovations in his computer design was putting in a keyboard so that the astronauts could enter verbs and nouns into the computer. It was the first ever small digital computer with a keyboard.) However, if the Primary Guidance, Navigation and Control System failed to operate anytime during a mission, the Block II Stabilization and Control System was designed to replace the primary system. The Block II Stabilization and Control System was a manual-controlled system and required the astronauts to fly the capsule. Besides being the backup system for the Primary Guidance, Navigation and Control System in the Block II capsule, the Block II Stabilization and Control System provided the Primary Guidance, Navigation and Control System with several full-time services while the primary system was in control of the Apollo flight. These services provided to the Primary Guidance, Navigation and Control System by the Block II Stabilization and Control System were required for normal Command Module operation. The Block I capsules, by contrast, did not have a Primary Guidance, Navigation and Control System and used the Block I Stabilization and Control System as the primary source of controlling the Apollo capsule maneuvers. Because the Block I Stabilization and Control System was the primary capsule control system on the Block I Command Modules and the Block II Stabilization and Control System was a backup control system in the Block II capsule, the Command Modules were different in their internal design. Now you know a few things about the Block II Stabilization and Control System, so let us get back to my family's move to Minneapolis.

The company had provided me with the date when the packers and movers were coming to move our furniture out of the house. This meant that our family had to leave for Minneapolis the day after the movers arrived, and since the company allowed seven days' travel time, I had to be in Minneapolis seven days later. The packers and movers came to the house to pack the furniture and other belongings on August 11 and began transporting them to Minneapolis. Since our house had not yet sold, we left it in

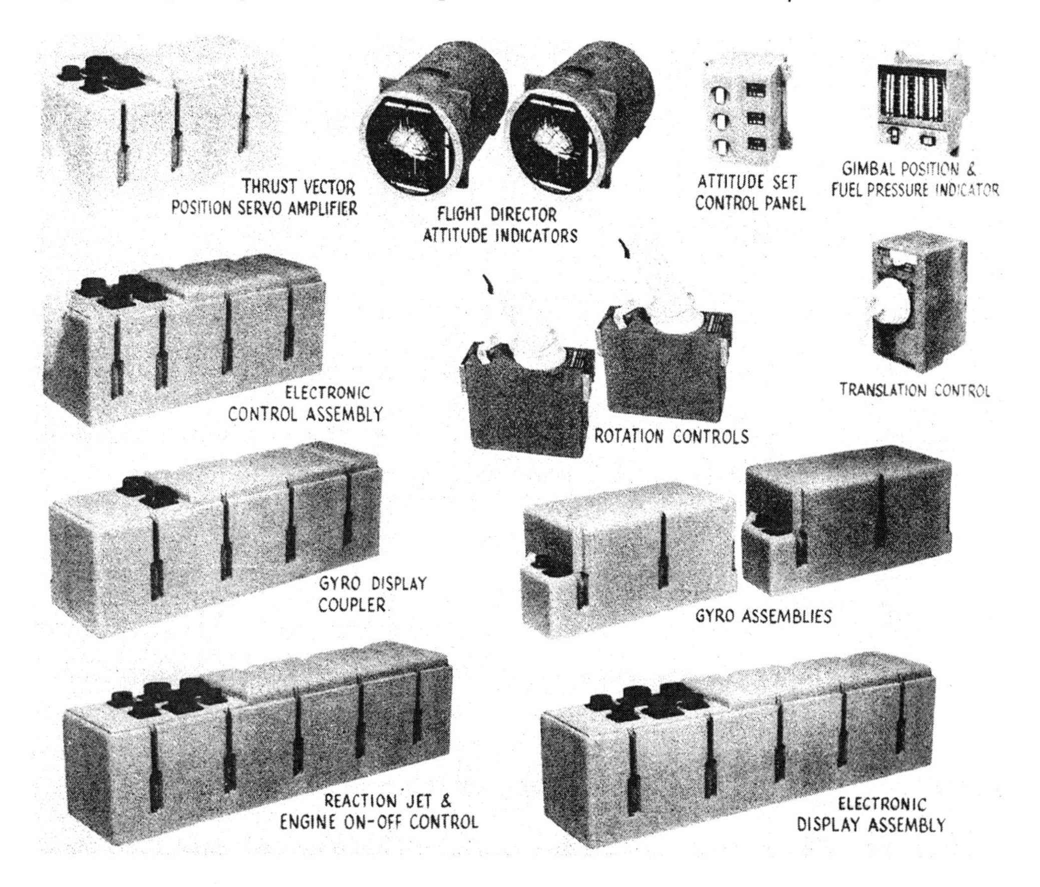

This is a photograph of the Block II Apollo moon mission Stabilization and Control System (Courtesy of the Boeing Company).

the care of the realtor, hoping that it would soon sell. That night we slept on the living room floor, with quilts and blankets spread out for everyone, and on the following day, August 12, our family was on the way to Minneapolis. We traveled up through Las Vegas and Salt Lake City, up the Snake River to the Grand Tetons and Yellowstone National Park, across Nebraska and Iowa, and then on to Minneapolis. After arriving, we lived in motels until we finally found a split-level home to lease for one year in the suburb of Fridley. The house was located in the Sylvan Hills of Fridley and belonged to Gordon and Dolores Olsen, who had moved to La Porte City, Iowa, to run a newspaper named *The Progress Review*.

After reporting to the North American Aviation subcontractor management office at Honeywell on Ridgway Road, Quality Control Rep Marv Smith, Quality Control Rep Ted Nakashima and I talked for a long time. They showed me their sleeping bags and said that during winter blizzards they usually slept in the office because there was no way they could drive home with one foot of snow on the roads. Right there and then, I decided that there was no way I was going to sleep in the office during a blizzard. I told Ted, Marv, Selby, and Jeannie that as soon as it was announced that a blizzard was heading our way, I was going to head for home, no matter what time of the day it was, because I was not about to leave my family to experience living through a snowstorm all by themselves. I was true to my word: all through the winter months I watched every blizzard on TV in the warmth of my home with my family in Fridley.

On August 25, 1966, five days after arriving in Minneapolis, AFRM 011 (the second unmanned Apollo capsule) was launched successfully on a suborbital ballistic trajectory. It was the first unmanned Apollo capsule with a Mission Control Programmer controlling the flight in place of an astronaut. The AFRM 011 mission tested the restart capability of the Apollo Service Propulsion System engine, which was extremely critical to the success of the Apollo moon missions. The Service Propulsion System engine was one of the engines that had to fire to make one (or possibly two) mid-course corrections to ensure that the future Apollo 11 spaceship was headed for the moon, fire again to slow down the spacecraft so it would be captured by moon's gravity and go into lunar elliptical orbit, fire once more to circularize the elliptical orbit and then again so that Apollo 11 could head back to Earth from the moon orbit with another firing to make a mid-course correction so that the Command Module would stay within the corridor and be captured by Earth's gravity for Earth reentry. The other section of the Service Propulsion System engine that was tested was the magnetic clutch and braking servo system that controlled the position of the engine nozzle (which was the area of expertise of Eagle Knight George Cortes). The Apollo ARFRM 011 mission should also make everyone proud of Paul Garcia, Larry Pivar, Danny Moreno, Bob E. Lee, George Cortes, Bill Cavanaugh and Gordon McLennan because they had helped design the Block I Stabilization and Control System that controlled the movements of the spacecraft throughout the mission and especially during the rolling entry maneuvers during Earth reentry by executing commands received from the Mission Control Programmer. After Earth reentry, the unmanned capsule was recovered 800 kilometers southeast of Wake Island in the Pacific Ocean. (Apollo Spacecraft 011 mission details can be found on the web on Apollo program and is listed as Apollo Mission AS-202.)

However, no success is ever attained with some sort of failure in the process, and this held true for AFRM 011. On July 19, 1966, a little over a month before I assumed the new engineer resident representative assignment at Honeywell, and while AFRM 011 was undergoing its pre-launch checkout on the pad at Kennedy Space Center, the Block I Stabilization and Control System suffered a significant anomaly in one of the electronic boxes. This was a major setback for Spacecraft 011 because if this fault were not corrected in time, the scheduled launch date of August 25 would have to be delayed. The Kennedy Space Center engineers removed the electronic devices that were suspected of causing the anomaly from Spacecraft 011 and had them tested. The tests verified that there was indeed a failure responsible for the anomaly. The electronic devices were returned to Honeywell in Minneapolis so that the engineers there could determine the cause of the problem and implement a fix. This problem put much pressure on Honeywell and its engineering staff because they had to find a solution before the scheduled launch date. During any spacecraft launch, no engineer or manager responsible for a system on board the spacecraft, much less any company that has a system on board, wants a failure of their system to be responsible for a delay in the launch date. Honeywell and its engineers came through like real champs because by early August they had characterized the failure sufficiently to come up with a repair to the Spacecraft 011 hardware that was agreeable to North American and NASA. The fixed hardware was returned to the Kennedy Space Center to be reinstalled in AFRM 011 without having to reschedule the launch. Honeywell engineers' response to quelling this panic tells you much about the type of people who were working on Apollo at Honeywell. While I was working on the Apollo program, each time a panic occurred, everyone swung into action and moved their asses to solve the problem as expeditiously as possible.

Regarding my new assignment, the question that came to my mind immediately after arriving at the office in Honeywell was this: What was going to be my main method of operation in this job? There were several options for handling problems as they occurred, and the choice was left up to me. During the two-week visit to Honeywell in July, I had spoken to Wes Jensen on how he handled the job. Wes told me that whenever an engineering problem arose in which a decision had to be made by him, he would defer a decision for later, telling Honeywell that he would have to confer with the engineers in Downey. He would then call the responsible engineer in Downey, lay the problem on him, and wait for a phone call from Downey with a decision. At the time that I was introduced to the Honeywell management, it occurred to me that Wes's tendency to delay making decisions by letting the Downey engineers do so was what prompted the Honeywell management to question whether I had the authority to make company-binding decisions. Also, recall that Ed Kelley had told Honeywell that I had the authority to make any engineering decision without consulting Downey and that any decisions I made would be binding. Of the two options, I favored making my own engineering decisions on the spot instead of calling Downey first. Calling Downey and waiting for a reply would only cause a delay in the program at Honeywell, and, anyway, it was not the way I was used to operating. The saying that it is better to ask for forgiveness than to ask for permission still stood for me. I was used to being in the middle of the action and was not afraid to make engineering decisions first and then argue why the decision I made was the right one. When you are put in a position of responsibility and a degree of authority, no matter how small, you must be able to make decisions with confidence and be ready to defend your choices; if you cannot do that, you do not belong there.

Remember that I was only my company's engineering resident representative at Honeywell and not the engineer responsible for the design of any part of the Block II

Stabilization and Control System as I had been on the Control Programmer and the Mission Control Programmer. The Block II Stabilization and Control System consisted of seven electronic boxes and seven displays and control units. The Downey-assigned responsible engineers for the design of this system were Lloyd Campbell (for the displays and controls) and Mert Stiles (for the electronic boxes). Also responsible for the test of the Block II Stabilization and Control System was Paul Garcia, with Bernice Johnston handling the qualification testing. These were four premier engineers with whom I would have to argue to defend the engineering validity of any decisions I made without consulting them first. Other engineers who would no doubt be involved in the Downey decision-making process were George Cortes and Carl Conrad. Carl was the Block II Stabilization and Control System Design Unit supervisor, and George was the Service Propulsion System engine clutch and servo system expert. Any time Honeywell requested a change to circuits or test procedures that involved the firing or controlling the Service Propulsion System engine, Eagle Knight George Cortes had to be involved or consulted.

At Honeywell, the project engineer responsible for the Block II Stabilization and Control System design was Chuck Moosbrugger, while Vern Johnson was the project engineer for the system's displays and controls. The Honeywell engineers who frequently came into my office to plead the Honeywell case for a system design change or a change of test tolerances were Dave Wilson, Fozzi Qutob, Eldon Lippo, Don Barnhill, and Don Carlson. Of course, after I gave Honeywell my decision, I would call Downey and talk to the responsible engineer there and anyone else he wanted to include in the telephone call. I would describe the system problem to the Downey engineers and tell them the decision that I had made, complete with all the pros and cons and the rationale behind my choice. There would follow a discussion about all the ramifications of the problem until everyone was satisfied that all was okay. I was also careful to tell Honeywell that the decision I made was final unless someone in Downey had strong objections, which left me an out to revise or reverse the decision in case someone in Downey wanted to make a change. During the time that I was the engineering resident representative at Honeywell, no one in Downey ever asked me to change any decisions I made, although there were many good arguments and hot discussions in which I had to aggressively defend my decisions until everyone in Downey was convinced that it was the right choice.

One of the important technical projects that Pete Smith of Honeywell had completed was performing and documenting a complex mathematical and thorough tolerance analysis of the Block II Stabilization and Control System. The tolerance analysis document was incredibly important because it allowed me to intelligently evaluate any Honeywell-requested Block II Stabilization and Control System test requirement changes, so they could be either approved or disapproved. It was mandatory that each system electronic box pass a test before it was delivered to the teams in Downey, and invariably Honeywell experienced problems passing the tests. Many times, Dave Wilson would come into my office to request changes to Block II Stabilization and Control System test specifications and used Pete's tolerance analysis to justify the modification. If the requested changes made sound engineering sense, were consistent with my understanding of the tolerance analysis document, and did not disrupt the operation of the Block II Stabilization and Control System, the request would usually be approved, but a phone call to Paul Garcia, Mert Stiles and others would follow to inform them of the change.

A few words here about Don Carlson are appropriate, because he was one of the friendliest Honeywell project engineers that I had the privilege of working with. Once a week, Honeywell would have a coordination teleconference with the engineers in Downey to discuss the problems that Honeywell was experiencing on the Block II Stabilization and Control System. A day or two before the telephone call, Honeywell would have an internal pre-conference meeting at which they previewed what they were going to discuss with the Downey people. I knew this because Don told me so. He also told me that sometimes there were problems that Honeywell did not want to discuss with the Downey engineers, so they would not put those problems on the agenda. However, Don said that while they were all told not to bring up these excluded problems, if someone from North American S&ID brought them up, they should not lie about them. I do not know how many such problems went undisclosed, but invariably some problems bothered Don, so he would come to my office and tell me about them and then would say, "I'm not going to bring up the problem, but if you bring it up, we can't lie about it." You can rest assured that every problem Don Carlson revealed to me got exposure during those telephone calls.

By September 6, my family's house in Santa Fe Springs, California, was sold. The papers to finalize the sale came by mail and were signed to finally transfer the house to the new owners. The house we leased from Gordon Olsen was ready for occupancy on September 14, and the furniture and other belongings were delivered on that date at my request. In late October, I received the formal paperwork for my promotion to resident representative, dated October 23. It was disappointing, extremely disappointing, because I was expecting the salary increase to be substantial, but it was not and did not make the move to Minneapolis worthwhile. The same week that I received the paperwork, I had a telephone call from Bill Fouts apologizing for the small increase. Bill told me they would make it up during the next review, but we all know that never happens.

In November 1966, just before Thanksgiving, it started to snow and did not stop for the next eight months. Minneapolis was buried in snow, and whenever a blizzard started, I headed for home, as I had promised, to be with my family, opting not to spend the night at the plant as other people often did.

15

Fulfilling the Honeywell Commitment and More Eagle Knights

There were still some problems to solve on the Block I Stabilization and Control System, so I dug into that task and managed to tie up all the loose ends. Apollo 1, using the Block I Stabilization and Control System, was now undergoing tests at the Kennedy Space Center. Sometime around mid–January 1967, I received a phone call from the home office in Downey reporting that astronaut Major Givens was flying in from the Cape with two Rotation Hand Controllers and two Translation Hand Controllers from Apollo 1 to be repaired. There would be a letter following Major Givens's arrival authorizing Honeywell to do the repair and instructing them on the amount of work to be done. I was to follow the four devices through the Honeywell process to ensure that all work and testing was done to Apollo engineering and manufacturing standards. One of the annoying problems with the Rotation and Translation Hand Controllers was that they were installed on or near the crew couches, with their cables hanging in areas where crew and maintenance personnel continually rubbed against them or stepped on them, causing damage to the cable covers. As a result, the cables and other parts of the controllers inevitably needed repairs. Later, I received a message that Major Givens was not coming but was never told the reason for the cancellation.

The original Apollo program plan for lunar flights was to go to the moon in the Block I capsules, with the Block I Stabilization and Control System serving as the primary method of guiding, navigating, and maneuvering the capsule. However, the Block I Stabilization and Control System had a low reliability for lunar flights, which meant that during a long lunar flight, the system, being analog and not digital, was going to have some electronic component failures, which could well be life-threatening to the crew. There was a mandatory plan to compensate for the low system reliability, but this plan turned out to be too cumbersome to implement and was troublesome for the crew. It hinged on the astronauts' ability to recognize life-threatening Block I Stabilization and Control System failures in specific system electronic boxes; they would then have to open the failed box and replace the failed electronic card with a good card that was carried on board with many other cards that contained different system functions. Because this approach was unworkable, a better concept had to be devised, and thus Block II was born.

The Block II concept was to replace the Block I Stabilization and Control System with a new primary system featuring a digital computer that would have all the application software programs to automatically perform the guidance, navigation and control tasks of the moon missions. The contract to design this new primary system was given to the Stark Draper Laboratories of MIT in 1964. (This is when the task of designing the digital computer for the new primary system was given to Eagle Knight Dr. Ramon Alonso.) A decision was also made at this time to redesign the Block I Stabilization and Control System as a backup system to the primary system and designate it as the Block II Stabilization and Control System. Although the Block I Stabilization and Control System was not sufficiently reliable for moon missions, it proved to be good enough for Earth orbital missions (which are relatively short compared to an extended lunar mission), which is why Apollo 1 was planned with a Block I capsule, as were the unmanned capsules, AFRMs 009, 011, 017 and 020.[1]

There were other shortcomings (known as constraints) in the Block I capsules regarding the Stabilization and Control System that made Block I unfavorable for moon missions. Sixteen operational constraints had been extensively discussed by Honeywell and North American Aviation and were well understood by both parties. These constraints had been documented formally in a letter from Honeywell to North American Aviation (numbered 6-Y-P-497), dated October 5, 1966, a copy of which is in my possession. The first operational constraint involved system performance during the critical Earth reentry flight phase; for the sake of the crew's safety, it was mandatory that the Block I Stabilization and Control System maneuver the Apollo capsule at peak performance, but the Block I system only performed marginally well, thus making the capsule more susceptible to being negatively affected by small errors. The second operational constraint was a system limitation that could cause the crew to lose awareness of the capsule's attitude during critical mission maneuvers, and the third operational constraint was a capsule power limitation exhibited under specific system switch positions that would inadvertently deplete all the fuel for the capsule's reaction jets, leading to certain disaster for the crew. Because the Block I Stabilization and Control System was intended for Earth orbital missions and the Block II Stabilization and Control System was intended as a backup system for manned moon mission Apollo capsules, in which the lives of astronauts would be at stake, there was much care taken in the design and manufacturing of both systems at Honeywell. I attended many meetings to discuss system design status and problems and had Honeywell engineering staff coming in and out of my office day in and day out throughout the program to discuss and solve numerous engineering difficulties.

The first manned Apollo launch was set for no earlier than February 21, 1967. A *Skywriter* article reiterated that the prime crew for Apollo 1 was Gus Grissom, Ed White, and Roger Chaffee; the backup crew was Walter Schirra, Don Eisele and Walter Cunningham.[2] On the afternoon of January 27, 1967, I got home from work around 5:00 p.m., as I normally did on Friday afternoons, just in time for dinner. At about 6:30 p.m., our family settled down to watch TV; half an hour later, the phone rang. I got up to answer it, and it was my boss, Ed Kelley, from Downey. Ed informed me that at 6:31 p.m. that evening there had been a fire in the Apollo 1 capsule and the crew members—Grissom, White, and Chaffee—had died. At that time, the news had not yet been broken to the public because the astronauts' families had to be notified first. I was told by Ed that NASA was going to impound all Block I Stabilization and Control System records, files, and any hardware at Honeywell; I was to make sure it all went smoothly. The whole accident was going to be investigated thoroughly, and employees were to fully cooperate with NASA in that regard. All members of my family were deeply saddened by the deaths of the Apollo 1 crew, and there was certainly a lot of work ahead for S&ID employees involved in the program after the fire investigation was completed.

The fire inside the capsule of Apollo 1 was the worst accident to occur thus far in America's manned space program. Our family spent most of the night listening to the news to see whether we could get more news on the tragedy. During the weekend, Honeywell was buzzing with activity, and when I arrived at work on Monday, all files, records and Block I Stabilization and Control System hardware were in a special storage place and had been impounded by NASA. I verified that it was well documented on paper that Honeywell had followed NASA's instructions to the letter. Honeywell followed up by sending a wire to North American Aviation on January 29, 1967, listing all the Block I Stabilization and Control System hardware, files, records, and data that had been impounded by NASA.

On April 11, 1967, as reported by *Skywriter*, nearly two and a half months after the Apollo 1 fire, the North American Aviation CEO, Lee Atwood, appeared before the House of Representatives regarding the Apollo 1 fire. He made a statement to the NASA Oversight Subcommittee of the House on Science and Astronautics that appeared in the April 21, 1967, issue of *Skywriter*. Mr. Atwood was accompanied to Washington, D.C., by Dale D. Myers, Harrison Storms, Thomas C. McDermott, George Jeffs and John L. Hansel. Mr. Atwood described to the subcommittee in some detail North American Aviation's position in the Apollo program and North American Aviation's organizational and operational methods.

Following Lee Atwood, Dale Myers, vice president and Apollo program manager, addressed the subcommittee. In Dale's opening statement, he told the committee members that we at North American Aviation were shocked and grieved to learn of the fire in Spacecraft 012 and that North American Aviation had offered its complete support in determining the cause of the fire. Dale went on to say that while North American Aviation was supporting the Board of Inquiry, the company, in collaboration with NASA, participated in what was called "Project Action." For this activity, North American Aviation put a group together to work with the NASA Manned Spacecraft Center at Houston to develop solutions to problems indicated by the accident. Out of this joint activity came the recommendation for numerous changes, including redesigning the fast-opening hatch, plumbing and wiring improvements, non-metallic material reviews, and criteria revisions. Dale informed the subcommittee that implementation of these recommendations would be discussed by the NASA representatives who were to appear before this committee later. Dale then explained how difficult it was to design a complex spacecraft such as the Apollo capsule because of the many variables that were involved. The design technique used to create such a vehicle is called "systems engineering" and involves a series of trade-offs among the many requirements. Dale emphatically stated that crew safety criteria were of paramount consideration in all trade-offs conducted. He then enumerated the design procedures, design reviews and tests implemented in the program to ensure crew safety.

Dale also made preliminary comments on the Board of Inquiry report even though North American Aviation had limited time to review it. Dale commented on the following five problem areas that were part of the findings in the board's report: (1) problems with the environmental control system, (2) the problem of solder joints, (3) electrical wiring, (4) no design features for fire protection, and (5) installation of non-certified equipment items. The Apollo Command Module and Service Module problems discovered in "Project Action," as well as those problems that were listed in the Board of Inquiry Report discussed briefly by Dale Myers, took many months and resources to fix. (Some of these solutions will be discussed in later chapters.)

The appearance of our CEO Lee Atwood, along with several North American Aviation high-level executives, was a clear and strong indication that North American Aviation was committed to making sure that the Apollo Command and Service Modules were redesigned to prevent another accident like the Apollo 1 fire.[3]

It is worth going into some detail regarding how the Apollo 1 fire occurred: On January 27, 1967, Gus Grissom, Ed White, and Roger Chaffee were training for the first crewed Apollo flight, an Earth orbiting mission scheduled to be launched on February 21. The three astronauts were taking part in a "plugs-out" test, in which the Command Module was mounted on the Saturn IB booster on the launch pad just as it would be for the actual launch, but the Saturn IB was not fueled. The plan was to go through an entire countdown sequence.

At 1:00 p.m., the astronauts entered the capsule on Pad 34 to begin the test. Numerous minor problems cropped up, which delayed the test considerably; finally a failure in communications forced a hold in the countdown at 5:40 p.m. At 6:31, one of the astronauts (probably Chaffee) reported, "Fire, I smell fire." Two seconds later, Ed White was heard to say, "Fire in the cockpit." The fire spread throughout the cabin in a matter of seconds. The last crew communication ended 17 seconds after the start of the fire, followed by loss of all telemetry.

The hatch in the Block I capsule could only open inward and was held closed by latches that had to be operated by ratchets. The hatch was also held shut by the interior pressure, which was greater than outside atmospheric pressure and required venting of the Command Module before the hatch could be opened. It took at least 90 seconds to open the hatch under ideal conditions. Because the cabin had been filled with a pure oxygen atmosphere at normal pressure for the test and there had been many hours for the oxygen to permeate all the material in the cabin, the fire spread rapidly, and the astronauts had no chance to get the hatch open. Nearby technicians tried to reach the hatch but were repeatedly driven back by the heat and smoke. One of the engineers at the fire scene was pad leader Don Babbit, whom I had briefly met at Eglin Air Force Base in Fort Walton Beach, Florida, when I was there in 1961 with the Hound Dog cruise missile.[4] Don attempted to open the hatch, but the heat was too intense.

By the time the technicians and engineers succeeded in opening the hatch, it was roughly 5 minutes after the fire had started; by this time the astronauts had already perished (probably within the first 30 seconds) due to smoke inhalation and burns. Thus did the United States lose three fine astronauts: Gus Grissom was one of the original seven astronauts and had flown in Mercury and Gemini capsules; Ed White II had flown in a Gemini capsule and was America's first spacewalker; and Roger Chaffee was a rookie astronaut who was going on his first mission.

The Apollo program was put on hold while an exhaustive investigation of the accident was conducted. The conclusion was that the most likely cause was a spark from a short circuit in a bundle of wires that ran to the left and just in front of Grissom's seat. The large amount of flammable material in the cabin in the oxygen environment allowed the fire to start and spread quickly. Numerous changes were instigated in the program over the next year and a half, including designing a new hatch that opened outward and could be operated quickly, removing much of the flammable material and replacing it with self-extinguishing components, using a nitrogen-oxygen mixture at launch, and recording all changes and overseeing all modifications to the spacecraft design more rigorously.[5]

The mission, originally designated Apollo 204 but commonly referred to as Apollo 1, was officially assigned the name "Apollo 1" in honor of Grissom, White, and Chaffee. The first Saturn V launch (uncrewed) in November 1967 was designated Apollo 4 (no missions were ever designated Apollo 2 or 3). The Apollo 1 Command Module capsule 012 was impounded and studied after the accident and then locked away in a storage facility at NASA Langley Research Center. The changes made to the Apollo Command Module following the tragedy culminated in a reliable craft that (with exception of Apollo 13) helped make the complex and dangerous trip to the moon almost commonplace. The eventual success of the Apollo program was a tribute to Gus Grissom, Ed White, and Roger Chafee, three brave men whose tragic loss was not in vain. So now let us return to the events taking place during the Honeywell assignment and review how North American Aviation proceeded to fix the problems that resulted from the Apollo 1 accident.

On May 1, 1967, Bill Bergen was appointed president of the Space and Information Systems Division, replacing Harrison Storms, who was on medical leave.[6] Bergen was a graduate of the Massachusetts Institute of Technology, had previously been vice president of the Space Propulsion Group at the company corporate office, and, before coming to North American Aviation in April 1967, was president of the Martin Company. The Space and Information Systems Division had also been renamed the Space Division. The company newspaper, *Skywriter*, issued two articles of interest on May 12, 1967. One article featured the introduction of Bastian (Buzz) Hello by Bill Bergen, division president, to the division's personnel at the Florida facility as the newly appointed vice president of launch operations. The second article featured a possible Reusable Space Transportation System that could reduce the cost of future manned space missions. No one paid much attention to the latter article, but this reusable system was the forerunner of the Space Shuttle, which would become one of the North American Aviation Space Division programs starting in 1969 (after Apollo 11 landed on the moon), the second largest space program for the division. During the week of May 12, 1967, the crews for the first Apollo manned mission were selected, with NASA astronauts Don Eisele, Walter Schirra, and Walter Cunningham serving as the prime crew and Tom Stafford, John Young, and Eugene Cernan as the backup crew. NASA, North American Aviation Space Division and the country were earnestly preparing for a moon mission.

Other indications that Apollo was moving toward a moonshot by the end of the decade were notices in the company newspaper in which employees were commended for doing great work that would get the Apollo program closer to a moon launch. There were some other Eagle Knights working in the Apollo program behind the scenes who were not engineers but still contributed on completing some critical work to the machines that were designed to take men to the moon and returned them home safely. Some of these Eagle Knights' names were published in the company paper and are featured below.

The following individuals were mentioned by name in the May 5, 1967, *Skywriter*: C.R. Lerma was noted as one of six graduates of White Sands test facility home study course on electronics technology. I never met Mr. Lerma, but I would speculate that the many hours he spent at home studying to graduate from the course allowed him to become an electronic technician and contribute to the Apollo program to further his technical career so he could earn a better living for his family. I say, kudos to Eagle Knight Mr. Lerma. Ray Tirado, Jose Lucero, Arcadio Bedolla, and Edward Elzondo, all of Apollo Bonding and Processing, were members of a team that completed no-defect primary bonding operations on Spacecraft 109 (Apollo 13). These four Eagle Knights did some fine work on the famous Apollo 13 spacecraft. Tom Chavez of Apollo Assembly Requirements was one of two employees who demonstrated the Apollo Assembly Requirements activities to Paul Vogt, newly appointed vice president and assistant to the division president. In addition, the following name was published in the May 19, 1967, *Skywriter*: "Diego Ortiz of Saturn S-II received his 20-year service pin."

By the beginning of June 1967, the Block II Stabilization and Control System design had already been completed and all major design problems had been resolved. What remained was incorporating design changes resulting from the Apollo 1 fire investigation and solving whatever problems were discovered during the manufacturing process. As I read one of the June issues of the company newspaper, it made me feel ready to return to California. The following names published in the June 16, 1967, *Skywriter* attracted my attention because those mentioned were some new Eagle Knights: "Ambrose Garcia, E.L. Cazares who work on the Apollo program and J.D. Azar who works on the Saturn S-II program have received their 20-year service pin." I cannot let this point go without a comment. At this time, these three employees had already worked for the company for 20 years; just to put this into perspective, I had worked for the company for 7 years, and my total experience in the aerospace industry only totaled 17 years.

Effective September 1, 1967, North American Aviation merged with Rockwell Standard to become the North American Rockwell Corporation, and it was now a larger company. The merger gave the company more resources to be able to capture and process more post-Apollo contracts, such as the Space Shuttle contract.[7] My year as resident representative at Honeywell would be over in August, but I was not sure I was going back to Downey because for the previous several months, on almost every telephone call with Ed Kelley, Ed kept hinting to me that it would sure be nice if NA Rockwell had someone at Honeywell for another year to handle the remaining problems. I was not ready to open the subject for discussion for fear that once I did, I would have to remain at Honeywell another year, so I ignored Ed and never acknowledged his statement. I was sure I did not want to take the bait, knowing that I had promised my wife that the assignment would be no longer than one year; however, I was worried about what would happen to my career if I did not volunteer to stay in Minneapolis. I struggled with this problem every hour of the day, knowing that I would eventually have to come to a decision even if it meant a career-limiting choice. I decided not to tell anyone about my predicament. I would just keep ignoring Ed's bait and see whether I could stand fast to my initial decision.

16

Returning to Downey and More Eagle Knights

Ed Kelley never asked me directly to remain at Honeywell beyond the agreed one-year period other than the numerous dropped hints. So, in one of our telephone calls, I told Ed to get the paperwork started for my relocation back to California. Ed said that he would let Bill Fouts know that I was coming back to Downey. As the time to leave Minneapolis approached, I wrote a letter to upper management outlining several optional methods for supporting Honeywell without a resident engineer. The lease on our Minnesota house was to expire in September 1967, and the Olsens wanted the house back because they wanted to sell it. The house had been for sale the summer before we moved in, but it had not sold, which was why they had agreed to lease it. Now they wanted to put the house back on the market. I informed Gordon and Dolores Olsen by letter that we were moving out sometime in August. Later, we received a letter from the Olsens dated July 19 thanking us for the prompt rent payments and for cooperating with the real estate agent and wishing us luck on our transfer.

The paperwork for my return to Downey was typed on July 20, 1967, to be effective on August 20, 1967. Thus I would have to report to work in Downey on Monday, August 21. The packers and movers came to the house on Trinity Drive on August 10 and emptied it all out, so that night we once again slept on the floor with quilts and blankets. At 4:00 a.m. the following morning, my family and I were on our way back to California. Leaving so early in the morning was not that difficult because I lowered all the rear seats in our four-year-old Rambler station wagon and made a bed for the kids in the back. Our decision this time was to take the most northerly route possible and follow the Missouri River to experience some of what Lewis and Clark saw on their wilderness trek. That route led us to travel across Idaho, Montana, Washington State, Vancouver, Victoria Island, Oregon, and California. It was a race to get to the Santa Fe Springs/Norwalk, California, area as soon as possible because come Monday, August 21, I had to report to work. On Saturday, August 19, I arrived in Santa Fe Springs.

I got a motel for the family and began to make plans to settle back in the area. I reported to work the following Monday and was assigned to work in the Stabilization and Control Unit headed by Carl Conrad. Aside from Carl and myself, the unit consisted of ten members: Jerri Sutton (secretary), George Cortes, Bernice W. Johnston, Danny Moreno, Sam J. Nalbandian, Jim E. Roberts, Tak Shimizu, Mert T. Stiles, Bill Van Valkenburg, and Bob O. Zermuelen.[1] The same week that I reported to work in Downey, the Customer Acceptance Readiness Review for AFRM 020 with the Mission Control Programmer was being prepared for delivery to NASA.[2] AFRM 020 was the last of the three unmanned Apollo Block I capsules with the Mission Control Programmer. The mission of AFRM 020 would qualify many Apollo items for moon missions, some of them being the Saturn V, Saturn II, and Saturn IVB rockets, the heat shield, and the guidance and navigation for accurate Earth reentry. The company newspaper issued on Friday, August 25, featured articles that again indicated that not only were there Hispanic engineers in my department, but there were also Hispanics in other areas of the division. For example, there were many Hispanics in the manufacturing area who were responsible for assembling the Apollo capsules and the Saturn II boosters, several of whom were featured in the company newspaper. Eagle Knights were present in the Guidance, Navigation and Control Department, as well as other places in the plant. This Eagle Knight's name was obtained from the August 25, 1967, *Skywriter*: S.I. (Jose) Jimenez from Apollo Logistics Training was a member of the North American Rockwell Team that went around the U.S. cities briefing people on Apollo 4 also known as AFRM 017 of unmanned Apollo that would fly its mission with the Mission Control Programmer. He would make his audiences aware of AFRM 017's importance in its launch and mission in America's quest for landing men on the moon before the end of the decade.

Before discussing my new Apollo assignment, a few words about the Guidance and Control Department, where I was assigned to work, are needed: As described earlier, I came to work on the Apollo program from the Hound Dog cruise missile program in 1964 and was hired by Bill Fouts, who was chief of the Automated Systems for the unmanned Apollo capsules. Bill was responsible for two units, the Automated System Unit and the Automated Design Unit; I ended up in the Design Unit working on the Control Programmer for AFRM 009. Bill Fouts' jurisdiction fell under the Guidance and Control Department, then being managed by Joe Dyson. Sometime before August 1, 1965, Joe appointed Bill assistant manager of the department. Sometime in 1966, just before I left for Minneapolis to serve as the Stabilization and Control System engineering resident representative, Joe Dyson quit the company to go into business for himself, and Bill was promoted to manager of the Guidance and Control Department. When I returned from Minneapolis in August 1967, Bill was still manager of the department.

Back in 1965, when the Apollo program was in full swing, the Guidance and Control Department had 137 employees (as detailed on the organization chart released on August 1, 1965). At that time, I was in the Automated Systems Design Unit working on the Control Programmer for the Unmanned Apollo capsule under Bill Paxton. The 137 employees in the Guidance and Control Department were distributed among eight units, each headed by a supervisor. There were also seven secretaries, one for each unit (in one case, two units shared a single secretary). In addition, Joe Dyson had four engineers on his staff, two resident representatives at Honeywell in Minneapolis (W.D. De Viney and W.D. Jensen), and one resident representative at MIT (G.L. Holdridge) for the Guidance and Navigation System. Joe also had two systems engineers—one for the Guidance System, R.A. Kennedy, and one for the Stabilization and Control System, R.E. Antletz, each with his own secretary.

The systems engineers on the Apollo program were extremely powerful people. Each system engineer was responsible for the technical performance of their respective systems, and any major design change had to be reviewed and approved by them before it could be approved for presentation to NASA. The final system design change approval, of course, had to come from the department manager, and he would sign off on the change only after the system engineer had reviewed and approved it. Whenever a major design change was made to any Apollo system, it had to be reviewed by responsible engineers, managers and project engineers of all other Apollo systems to assess any impact that the change would have on their system. Additionally, all major design changes had to be approved by the NASA systems engineer, who, for the Stabilization and Control System, was Orville Littleton from the Manned Spacecraft Center in Houston. Orville would always include his boss, Gene Rice, in the decision-making process. This all took a tremendous amount of coordination, in meetings, between design units, responsible engineers and NASA. Because of the national prestige associated with going to the moon before the end of the decade, the cost of any major Apollo system design change made prior to July 1969 that would enhance the probability of achieving a moon voyage (or any major design change that was made to enhance the safety of the astronauts) was never seriously challenged.

So, where is all this background information leading? Through facts gleaned from organization charts, what follows are some statistics demonstrating how in the 1960s the engineering profession was controlled and ruled by white male engineers with and without engineering degrees, as well as how women and minorities fared in the profession. The lack of diversity that existed during this era will not be surprising. The discussion will be centered on the Guidance and Control Department of the Apollo program, but this department was typical of the hundreds of companies and departments that it took to build a spacecraft that could transport three astronauts to the moon and back to Earth safely. The same situation prevailed when I worked for Hughes from 1954 to 1960.

Of the eight units under Joe Dyson on August 1, 1965, only one—the Simulation Requirements Unit—was supervised by a female, B.S. Siev. She had thirteen engineers in her unit, only one of whom was Hispanic, by the name of D. Hernandez (no doubt another Eagle Knight). The Stabilization and Control System Design-Block I Unit was supervised by my Hound Dog cruise missile mentor and supervisor Ed Kelley, and it consisted of sixteen engineers, two of them Hispanic—George Cortes and Paul Garcia—and one female—Bernice Johnston, who had a degree in math. (The reason the math degree is noted is that during this period very few women had engineering degrees.) The Stabilization and Control System Design-Block II Unit was supervised by Don Niemand; of the twelve engineers in his unit, two were Hispanic—Danny Moreno and Pete Ontiveros. The Guidance and Control Systems Unit, supervised by Larry Hogan, had no minority or women engineers. The Guidance and Control Laboratory Unit was supervised by John Notti, with whom I had worked on the Hound Dog cruise missile program, and of the eighteen engineers working for John, one was Hispanic—Cruz Mora—and two were Asian—Sam Okada and Roy Tomooka. The Automated Systems Design Unit, supervised by Bill Paxton, had fourteen engineers; I was the only Hispanic, serving as lead engineer because I was the only engineer with enough experience to do the job. The Guidance and Navigation Installation Unit, supervised by Al Zeitlin, had twelve engineers, with one Asian—Chuck Yee—and no females. The Entry Monitor Systems Design Unit was supervised by Dale Bennett, whom I knew from the Hound Dog cruise missile days when Dale worked for Autonetics. Dale supervised seven engineers, none of whom were members of a minority ethnic group or female. Obviously, it can be seen from this description that there were no unit supervisors who were members of a minority group. As far as I can remember about the Apollo organization during this period, I was one of very few Hispanics in a lead engineer position with a reasonable chance of being promoted to supervisor, but given the state of diversity and the prevailing attitude toward women and minorities, coupled with company culture in engineering in corporate America, it was not going to happen without some outside pressure, and I knew it. Other Hispanic Eagle Knights who could have been in a lead position at the time were Marty Gutierrez and Bob Loya from the Electrical Power Department.

Each of the units under Joe Dyson, by charter, served a specific purpose in the Guidance and Control Department. The Stabilization and Control System Design–Block I Unit was responsible for the design of the Block I Stabilization and Control System and oversaw Honeywell to ensure that they were mechanizing the Stabilization and Control System for the Block I Apollo capsules according to the specification imposed on them by the design unit. The Stabilization and Control System Design–Block II Unit performed a similar function, except that its responsibility was the Stabilization and Control System that was to be the backup system to the Block II Primary Guidance, Navigation and Control System for the moon-rated Block II Apollo capsules. The Guidance and Control Systems Unit supervised by Larry Hogan provided system design requirements to all of Joe Dyson's design units for the Guidance and Control Systems. The engineers in Larry's unit would take the NASA top-level requirements for the systems and reduce them to individual Guidance and Control Systems' requirements, which were then relayed by written documentation to the individual system design units. The requirements that came from Larry's unit were usually in the form of logic diagrams and mathematical equations. The design units would in turn use the requirements from Larry's unit to design a system that would meet the NASA requirements. The design of each system was documented in a system specification and a work statement written by members of the design units that in turn were sent to the subcontractor who was to mechanize and manufacture the system in hardware form for delivery to North American Aviation.

In prior chapters, the responsibilities and functions performed by the Automated Systems Design Unit under Bill Paxton were documented, so that information will not be repeated here. The Guidance and Control Laboratory Unit supervised by John Notti served as the lab in which Guidance and Control Systems were tested to verify and certify various operations. That is where Eagle Knight Cruz Mora was working in 1965 when the AFRM 009 Control Programmer team was performing lab tests on the Bread Board Control Programmer. (In long conversations with Cruz, he told me that he had transferred from the Los Angeles Airplane Division and that he wanted to go back but there was no work for him there.) The lab provided test equipment, test facilities and engineers to perform each of the systems tests being performed in the lab. The unit responsible for the design of any system being tested in the Guidance and Control Lab wrote the detailed test procedures for their system, requested that their test be scheduled to be run in the lab and provided design engineers to support the tests being performed on their system by the lab engineers. If any anomalies were discovered during its test, it was the responsibility of the engineers from the design unit, in conjunction with the Guidance and Control Lab engineers, to find the reason for the anomaly. Once the cause was found, the design engineering team would have to make a design change to the system to purge the anomaly or to change the test procedure if it was found that it was in error.

The Guidance and Navigation Installation Unit under the supervision of Al Zeitlin was responsible for issuing drawings and process specifications that documented detailed instructions of how and where in the Apollo Command Module the Guidance and Navigation Systems hardware was to be installed. Believe me, this was no easy task. When a spaceship is being guided to the moon and back again, there is very little margin for error, so the installation of Command Module equipment used for guidance and navigation of the capsule had to be precise. B. Siev's unit, the Simulation Requirements Unit, did exactly what the unit's name implied: it provided requirements for computer simulations of the Guidance and Control Systems. The last of Joe Dyson's eight units, the Entry Monitor Unit, was responsible for the design of the Apollo Entry Monitor System, a critical system that needed to function properly in order for the capsule to reenter Earth's atmosphere (this system will be discussed in greater detail in chapter 22).

This is a good place to talk a bit about the individual Hispanic engineers in the Guidance and Control Department because, besides being premier engineers, they were Eagle Knights. Paul Garcia is a good one to start with because I met him when I was working on the Hound Dog cruise missiles from 1960 to 1964, back when diversity in corporate America was nonexistent, and Paul was one of only three Hispanic engineers I had met between 1954 and 1960. I had known Paul the longest, and yet I did not really know intimately about his personal life. Paul, so he told me, was a graduate of Garfield High School in East Los Angeles. He had been in the Navy during World War II, and one day during a conversation Paul showed me a picture he carried in his wallet of himself in his Navy uniform when he was young and as tall as a bean pole (Paul was over six feet tall). I believe he was a graduate of Northrop Institute. Paul had worked on the Navaho/X-10 program with Ed Kelley, Bob Antletz, Gary Osbon, and many other North American Aviation old-timers. Working on that program, as well as the Hound Dog cruise missile program, had given Paul a tremendous amount of experience and made him one of the best flight control systems test engineers at North American Aviation. Because of his long tenure at the company, Paul personally knew many of the executives because he had worked alongside of them or for some of them on their way up the corporate ladder. Years later, Gary Osbon, who had been chief engineer for the Hound Dog cruise missile program and later chief engineer for the Apollo program, became president of the FINCOR Division of Rockwell International, which dealt with South American countries in the sales and distribution of Rockwell International products. Paul told me one day that he had gotten a call from Gary offering him a position in the FINCOR Division, but he had thought it over very carefully and ultimately told Gary that he could not accept the position.

The second Hispanic engineer in the Guidance and Control Department was George Cortes. As mentioned in previous chapters, George was the expert in the servo and magnetic clutch arrangement that moved the Service Propulsion System engine nozzle into proper firing positions—a critical operation for firing the engine correctly. George recounted the story to us one day that he and his wife had bought a beautiful silky terrier puppy that they would leave alone in their home during the day while they were gone. One day when they returned home in the afternoon, they had to rush the terrier to the veterinarian to have his stomach pumped because the dog had eaten the lead weights at the bottom of the living room curtains. As George told us the story, it sounded funny, but at the time that it was taking place it must have been a traumatic episode. George was respected by everyone in the Apollo community, and whatever he said regarding his area of expertise was not challenged by anyone, not even members of management or NASA. On one occasion, as Bill Fouts recounted it, Bill was making a presentation to some of the NASA managers at the Johnson Space Center in Houston, Texas, regarding the operation of the system on which George Cortes was an expert, and Bill was asked a question concerning the system operation. Bill explained its operation exactly as George had explained it to him before Bill left for the meeting, but the NASA manager who asked the question wanted to know why the system operated that way. Bill did not know the answer and, after thinking for a few seconds, he said, "Because George Cortes says that's the way it operates." That answer brought laughter from the crowd and ended the questioning. Bob Antletz, who was systems engineer for the Stabilization and Control System, kept spelling George's name "Cortez"; despite George repeatedly telling Bob that his name ended with an "s," Bob kept spelling his name with a "z." Finally, to cure Bob of the error, George started to spell Bob's last name "Antlets" instead of "Antletz." After many, many times of George doing this, Bob got the message. This, of course, was all done in fun between the two of them over several months.

As for Pete Ontiveros, I knew little about him during the Apollo program, but in later years I learned that Pete was a very capable and excellent engineer. Pete worked in Guidance and Navigation, and he understood very well the mathematical significance as applied to that part of the system. From Pete I learned that he was a friend of Macario Garcia from Sugarland, Texas, just outside of Houston; Macario was a World War II survivor and recipient of the Medal of Honor, awarded by Congress, whose valor is chronicled in Raul Morin's book *Among the Valiant* (see pages 143 to 148 of this book for more information). When I was in the Air Force and stationed at Ellington Air Force Base outside of Houston, I was at a dance where someone pointed out Macario Garcia to me, but even though I saw him, I never personally met him.

This photograph illustrates the Block II Apollo Stabilization and Control System and the required documentation on delivery to North American Aviation by Honeywell to satisfy the NASA requirements (Courtesy of the Boeing Company).

I know very little about the remaining three Hispanic engineers: Cruz Mora, whom I have already mentioned; Danny Moreno, who was a younger engineer and no doubt hired into the Apollo program out of college; and D. Hernandez. But one thing is certain: any Hispanic who had earned an engineering degree in the 1940s, 1950s or 1960s had to be good because I know from experience that engineering degrees were not dished out like cereal. Another Hispanic and Eagle Knight I never met while working on the Apollo program was Dr. Ramon Alonso from MIT, who, as mentioned previously, was the designer of the digital computer that constituted the brains of the Apollo Primary Guidance, Navigation and Control System, as well as the digital computer in the Lunar Module that landed two astronauts on the moon and took them up safely to rendezvous with the main Apollo craft. This Control Module Computer contained an operating system like Microsoft Windows, a guidance application program, a Navigation Application program, a Control Application program, and other application programs, which now are known as apps (so do not think that apps were invented in the modern age for smartphones).[3]

While the design of the Apollo spacecraft in Downey was nearing completion and engineering staff was being reduced, the number of personnel at the Kennedy Space Center was increasing to be able to handle the upcoming traffic in Apollo moon missions. As the date for the moonshot approached, it took less and less system design engineering manpower in Downey, California, to achieve a moon mission by 1969. None among the Hispanic engineers knew how long we would be retained in the workforce, and so we had to take our chances along with the rest of the engineering staff and hope that the company would retain us.

17

A New Assignment and More Eagle Knights

Now that you know all about the Guidance and Control Department organization from the previous chapter, let us get back to my return to California from Minneapolis, which was followed by my new assignment. The first order of business for me was to get some rest and then search for a place to live. My wife and I started looking for a house to rent and eventually found one in Norwalk not far from the intersection of Pioneer Boulevard and Imperial Highway and just east of I-5. We moved into the house on September 9, 1967, with the address of 11928 Pioneer Boulevard.

The September 15, 1967, issue of the company newspaper, the *Skywriter*, announced that the merger between North American Aviation and Rockwell International was set for September 22. The company was now going to be bigger and better with more resources, able to capture more impressive contracts that would supposedly make employees' jobs more secure. Also, engineers were no longer going to be research engineers but members of the technical staff. Other sections of the same issue of the company newspaper listed some behind-the-scenes activities of various Eagle Knights, as follows:

> Tony Lopez of Apollo CSM Sub-Systems Assembly and Test is a member of the department that installed and tested the sub-systems that went into the Command and Service Module Assemblies of the Apollo Spacecraft.... Cecil Suarez is a member of the Central Quality and Reliability Assurance team that inspects all work of assembling and test processes performed on the Command and Service Module Assemblies for compliance with high quality and reliability requirements of the Apollo Program.... Rueben Ramirez is a member of the Plant Services organization that provides any services required to maintain the plant in operation.... Hank Lacayo is President of the UAW Local 887 Union.... A.J. Castillo of the Apollo Command Service Module (CSM) and Baltasar Diaz of Manufacturing and Facilities received 20-year service pins. [This achievement is impressive, as the math indicates that these two knights started working for the company in 1947, about two years after World War II ended] ... Ambrose Garcia and Joe Aragon of the Apollo Assembly Requirements organization are mechanics on a team of eight mechanics that performed perfect welds on Spacecraft 108 (Apollo 12) and 110 (Apollo 14) and only received three squawks (imperfections that were later corrected) on Spacecraft 107 (Apollo 11) and 109 (Apollo 13).

This last point was no small accomplishment because the Apollo capsule was composed of an inner frame that served as the astronauts' habitat and an outer frame that served as the structure to withstand all the forces the capsule experienced during the mission. There were hundreds of welds that had to be made to hold these frames together, and completing them without error like they did on Apollo 12 and Apollo 14 is phenomenal and fantastic, as is having only three imperfections on two famous Apollo capsules (Apollo 11 and Apollo 13; working on both of these capsules was a unique achievement that will never again be matched). The heat shield was what kept the astronauts from burning up during reentry, and, just to illustrate how critical those welds were, it was the outer structure of the capsule that kept the astronauts from being crushed to death from the dynamic pressures experienced by the capsule during the launch phase and the Earth reentry flight phase.

One day in October 1967, my wife came across an ad for a house for sale with no down payment. The real estate agent affirmed that there was no house loan down payment required, so I made an appointment to see the house and met the agent on the appointed date. The house had three bedrooms and two bathrooms, and it seemed to have been vacant for quite some time. The agent said that the realtor he worked for had ended up with it. The house had not been able to sell, so the realtor just wanted to get rid of it. The agent did say that the realtor wanted $100 upfront earnest money, and the buyers could move in as soon as it was cleaned and painted. While the deal was in escrow, the rent would be $110 a month, and the buyers would take possession thirty days after they moved in. He also stipulated that of the $110, $55 would be allocated for paint and cleanup. On November 11, my wife and I met the agent to close the deal. For $210 cash money, we would own a home, 16345 Denley Street in Hacienda Heights, and would be moving in almost immediately and take possession on December 15, 1967.

Two days before I met with the agent to close the deal on the house in Hacienda Heights, AFRM 017 was launched. AFRM 017, with its Mission Control Programmer in control, was launched on November 9, 1967, for two Earth orbits, at which time the S-IVB booster (called the S Four-B booster) and the Service Propulsion System engine fired to put AFRM 017 into a higher orbit. The Service Propulsion System engine then fired again, with the capsule pointing toward Earth, sending the spacecraft toward Earth at a higher speed than it would achieve on a return trip from the moon. The Command Module was subsequently separated from the Service Module, followed by the Mission Control Programmer commanding the Stabilization and Control System to fire the proper reaction jets in the Command Module to turn the module with the heat shield facing Earth for reentry. The initiations of Command Module–Service Module separation, engine and jet firings were all flawlessly orchestrated by the Mission Control Programmer. The higher reentry speed tested the Apollo heat shield at a hotter temperature than when returning from the moon and certified the heat shield for moon missions. The Apollo spacecraft carrying AFRM 017 was launched into space by a Saturn V rocket, marking the first time that this rocket was used to put a payload into space. (The Saturn V was the heavy-lift rocket booster developed by Wernher Von Braun's German team and was needed for the Apollo moon missions.) AFRM 017 was recovered near Hawaii. Details of the mission can be found on the web; it is listed as Apollo Mission AS-501 (or Apollo 4). The success of AFRM 017's mission was lauded as a major milestone by NASA's Dr. George Mueller and put the Apollo program one step closer to a possible moon mission by 1969.[1]

I was extremely proud of the success of AFRM 017 because my team and I had been instrumental in the design and performance of the Mission Control Programmer. What made me even prouder was that Paul Garcia, George Cortes, Danny Moreno, Pete Ontiveros, and Cruz Mora—five tremendous Hispanic engineers and Eagle Knights who were members of the Guidance and Control Department at the time—had been partially responsible for the performance and success of the AFRM 017 Block I Stabilization and Control System, which was no insignificant task. It was the system responsible for positioning the capsule and nozzle of the Service Propulsion System engine for each of the two Service Propulsion System engine firings, responsible for firing reaction jets to maneuver the Apollo space capsule into position during flight to ensure that the heat shield was pointing in the right direction for Earth reentry, responsible for firing the Command Module reaction jets for the rolling entry maneuvers, and responsible for firing the reaction jets that held the Apollo capsule stable throughout its flight prior to being captured by Earth's gravity. Of course, these feats were all accomplished by the Block I Stabilization and Control System responding to commands it received from the Mission Control Programmer.

An additional valuable piece of data recovered by the cameras mounted in AFRM 017 was a video of Earth that showed what the planet would look like as seen by the astronauts on their way to the moon during the first Apollo manned moon mission. The video of Earth and still photos developed from the video were circulated for view by the general public and were a big hit. The AFRM 017 Command Module was later put on display in Building 247 in the company's Downey facility for employees and their families to view. There were nearly 3,000 data points for review to evaluate the success of the mission that were returned from the space vehicle, and many of those data points helped to eventually determine that the Apollo heat shield had exceeded predictions, making this one of the most successful Apollo flights to date.

However, the road to the launch of Apollo 11 to the moon before the end of the decade was still not secured because AFRM 020, the last of the unmanned Apollo capsules, had to be successfully launched to prove that the capsule and its systems could withstand the cold in the vacuum of space experienced in a long-duration voyage to the moon and back to Earth. More important was that AFRM 020 was going into deep space and then returning to Earth at moon return velocity and through the moon return Earth entry corridor, which would prove the guidance and navigation technique for Earth reentry. More difficult still was the recovery from the Apollo 1 fire, which had claimed the lives of three astronauts. To be able to go to the moon, the Apollo capsule's ingress and egress hatch would have to be redesigned so that it could be quickly opened from inside by the astronauts, internal capsule equipment had to be purged of all flammable material and replaced with nonflammable material, and the capsule's 100 percent oxygen environment would have to be replaced with something other than pure oxygen, but that subject will be discussed soon enough.

On November 17, 1967, I received a Technology Utilization Award Certificate at work for a reduction of relay test points in the Mission Control Programmer for AFRMs 011, 017 and 020. This award was shared with Thom Brown for work completed on the redesign of the Mission Control Programmer test points while we were working on the unmanned Apollo capsules. The certificate reads, "T.U. No. SID 59165—REDUCTION OF TEST POINTS FOR RELAY LOGIC." The work listed in the T.U. certificate is one of my numerous contributions to the successful flight of AFRM 017 and the hopeful future success of AFRM 020. NASA was pushing North American Aviation to reduce the cost of the Apollo program, and so the company instituted different types of awards for various categories of savings, with monetary awards given for large money savers.

There was no clue as to what assignment Carl Conrad had for me on returning from Minneapolis, but it was not a surprise when, as lead engineer, I was asked to perform the same tasks that I was performing at Honeywell as the resident representative, in addition to being given responsibility for the design of the Block II Stabilization and Control System displays and controls (previously the responsibility of Lloyd Campbell, who was no longer in Carl's unit), with the additional task of purging flammable materials from the Block II Stabilization and Control System displays and controls in preparation for a moon flight. Fortunately, Carl was nice enough to assign Chuck Markley to assist me in purging flammable materials from the Block II Stabilization and Control System tasks.

To reaffirm the method of operation with Honeywell and to designate who in Downey could authorize Guidance and Control System design engineering document changes in the absence of a North American Aviation engineering representative from Downey at Honeywell, Ed Kelley, the Block II Stabilization and Control System assistant designated subsystem project manager, sent a letter to Wally Lundahl, Honeywell Block II Stabilization and Control System chief engineer, four weeks after my reporting to work from Minneapolis, stating that I, as primarily responsible (with Mert Stiles as alternate), had full authority to fulfill the responsibilities that were previously the purview of the engineering resident representative.[2] Basically, my assignment encompassed responsibility for the design for the Translation Hand Controller, the Rotation Hand Controller, the Attitude Set Control Panel, the Flight Director Attitude Indicator, and the Gimbal Position/Fuel Pressure Indicator of the Block II Stabilization and Control System for Apollo capsules manned missions. The Block II Stabilization and Control System was going to fly with the manned Apollo capsules starting with Apollo 7 and ending with Apollo 17. The design responsibility part of my assignment plus the resident engineering assignment required that I travel to Honeywell in Minneapolis repeatedly to help solve any design problems with any one of these devices and that I be at Honeywell during the implementation of any design change requested by the company or by NASA or whenever any engineering design and test document required review and approval signature.

The requests for me to travel to Honeywell came from several sources. Most of the time, the request came from Chuck Moosbrugger, the Honeywell Block II Stabilization and Control System project engineer. Chuck's request was generally prompted by his design engineers wanting me to be present at Honeywell for on-the-spot approval of a design or test document or to witness a test that required our company's approval or to approve some design change to the system. I would normally discuss problems regarding the Rotation Hand Controller with Steve Scarborough (who was responsible for the design of the Translation Hand Controller and the Rotation Hand Controller) and Eldon Lippo (responsible for the design of the Attitude Set Control Panel, the Flight Director Attitude Indicator, and the Gimbal Position/Fuel Pressure Indicator). If they felt that I should be present at Honeywell and I agreed that I should be there, they would ask Chuck to send a formal request that I be at Honeywell for whatever calendar period the Honeywell engineers felt was necessary.

The Block II Stabilization and Control System displays and controls were of relatively high importance to the Apollo missions during the time that the Command Modules were under the control of the Primary Guidance Navigation and Control System. Even though the Block II Stabilization and Control System was a backup system for the Primary Guidance Navigation and Control System on Apollo, the displays and controls of the Block II Stabilization and Control System that were under my jurisdiction were especially important because during the real time of any Apollo mission, any abort of the mission would be initiated by the astronaut turning the handle of the Translation Hand Controller in the counterclockwise direction. On possible failure of the Primary Guidance Navigation and Control System, the switchover from the failed Primary Guidance Navigation and Control System to the Block II Stabilization and Control System would be accomplished by turning the handle of the Translation Hand Controller clockwise.

In addition, some functions of the Stabilization and Control System Displays were active and primary throughout the Apollo mission. For example, during all Apollo flights, the crew monitored the attitude of the Apollo Command Module on the Flight Director Attitude Indicator, a unit that was under my jurisdiction. There were other Block II Stabilization and Control System functions that were used during a normal mission, but these are too numerous to list here.

Because originally the Apollo moon missions were going to take place in a Block I Command Module, with the Honeywell Block I Stabilization and Control System as the primary system that was going to control and navigate every Apollo capsule to the moon and back, the astronauts were involved in the design of the Block I Stabilization and Control System. Remember that at the beginning of the Apollo program in 1961, the Block I Apollo capsule required the astronauts to fly the capsule manually because the Block II capsule concept was not conceived until 1964. So, because of the close visual interaction the astronauts had with the displays during a mission and their active interaction with the Translation Hand Controller and the Rotation Hand Controllers, the Apollo astronauts had a very deep involvement at the beginning of the program in contributing to the design requirements of these displays and controls.

On my return from Minneapolis, I inherited several documents that described the early design requirements for some of the displays and controls. One of these documents

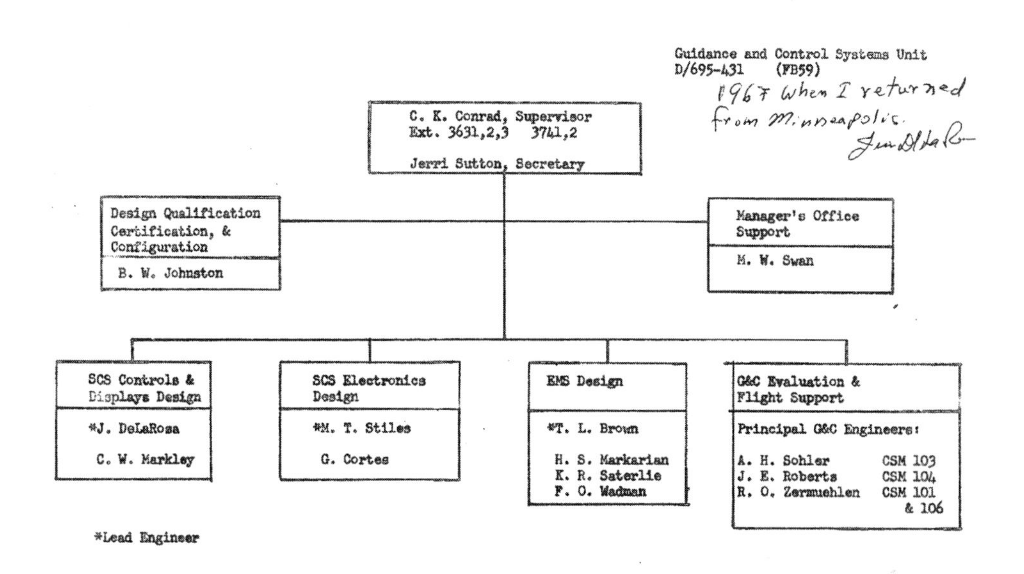

This is an enclosure to a North American Aviation internal letter issued in August 1967 formally announcing the Apollo Guidance and Control Systems Unit organization when the author had returned from Minneapolis (Courtesy of the Boeing Company).

was a Functional Specification for Preliminary Attitude Controller (the Attitude Controller was in essence the Rotation Hand Controller because this device manually controlled the attitude of the Apollo capsule by firing the reaction jets), with a revision date of September 10, 1964, which was prepared by Roy L. Cox and Edward H. White II (the first American astronaut spacewalker, who perished in the Apollo I fire) and approved by D.K. Slayton (one of the original seven astronauts), J.W. Bilodeau for W.J. North, A.B. Shepard (another of the original seven astronauts) and Donald C. Cheatham. The copy of this document belonged originally to Mort Fox (his name is printed on the cover page), who was long gone from the program. Another document from Mort Fox was a record of minutes of the Command Module crew integration systems meeting No. 1, dated September 10, 1964. Items discussed during this meeting were the Block II Apollo Capsule Mockup review, the location of the Rotation Hand Controllers in the astronauts' couches in the Block II Apollo Capsule Mockup, the control and display panel arrangement in the Block II Apollo capsule and the need for the astronauts to submit the revision of the Functional Specification for Preliminary Attitude Controller previously discussed. Attendees at the meeting were three engineers from North American Aviation (one of them being Mort Fox), two engineers from Honeywell, one engineer from General Electric, and fifteen personnel from the NASA Mission Control Center in Houston. Some of the NASA attendees were J.W. Bilodeau; Gene T. Rice, the NASA Stabilization and Control System subsystem manager; and astronauts Dick Gordon, Don Eisele, Charles Conrad, Jim McDivitt, and Ed White.

When I returned from Minneapolis, I already knew most of the engineers in Carl's unit from having worked with them before the Minneapolis assignment and from having worked with them over the telephone. One engineer in Carl's unit with whom I was only slightly acquainted was Bill Van Valkenburg, but I got to know him much better after my return from Minneapolis. Apollo was an exceedingly high-stress program because of the pressure everyone was under to get the design right and to maintain the schedule for a moon landing in 1969; consequently, many employees suffered from ulcers, hypertension, strokes, and heart attacks, from which some did not survive—there were even some suicides. To counteract the effects of the high stress, the company recommended and encouraged that all employees exercise on the gym equipment at the Recreation Center (which was open all day and up to late evenings) or walk or jog during the lunch break or after working. Bill Van Valkenburg and I went to the Recreation Center daily to jog, shower at the center and have lunch afterward. The closest I came to anyone experiencing a heart attack at work was when a group of paramedics came in and took my former supervisor Bill Paxton to the hospital in what turned out to be a simulated heart attack. In 1984, my former boss and mentor Ed Kelley passed away from a massive heart attack while working on the Space Shuttle program. I received the bad news on his passing on Monday, February 13, 1984, when I was no longer working on the Space Shuttle and was working in Sunnyvale as a consultant to the Aerospace Corporation of El Segundo, California.

My first trip to Honeywell in 1967 took place from October 11 to October 27. I was required to write a report detailing the accomplishments during each of my trips. This particular trip to Honeywell in October 1967 was for much work that was waiting for me because Honeywell had been without an engineering representative from Downey since my departure in late August. The purpose of the trip was to provide on-the-spot approval of Acceptance Test Requirements for Block II Stabilization and Control

Systems for Spacecraft 2TV-1 (a boilerplate spacecraft), Spacecraft 102 (Apollo 8) and subsequent spacecraft. However, there were so many problems that I did much more than was originally requested.

The trip's activities were documented in an internal letter dated November 17, 1967. The report included a narrative of my activities and five enclosures that documented the details of all the tasks accomplished while at Honeywell. I also attended a briefing given to high-level Honeywell executives to review the various hardware problems Honeywell was experiencing with the Apollo system. All major problems and their proposed solutions were discussed during the briefing. It was also in October 1967 that North American announced that they were building a new facility in Laguna Niguel, where the company had bought a large piece of property. The plan was to build the facility so that the Space Division could move into it, and, in addition, the company was going to build houses on the surrounding property for employees to buy so that all employees could walk to work and not have to commute by car. It sounded like a great plan, but it was eventually abandoned.

18

The Manned Apollo Capsules and More Eagle Knights

As we were approaching the end of 1967, Honeywell was already delivering Block II Stabilization and Control Systems for the Block II Apollo Command Modules. Some of my trips to Honeywell during 1968 were associated with the test of the Block II Stabilization and Control System, the design of the displays and controls, and the purging of flammable materials for the Apollo moon missions that would help recover from the Apollo 1 fire. I will relate these activities later.

In the January 12, 1968, *Skywriter*, it was reported that Eagle Knights Ray Tirado (a supervisor in the Apollo Bonding and Processing organization), Bob Cervantes, Victor Almanza, and Jose Gomez (all of Apollo Bonding and Processing) were members of the 35-man team that received personal congratulations from astronaut Walter Cunningham, who was representing Alan Shepard, chief of the NASA Manned Spacecraft Center's Astronaut Office, for completing the error-free and ahead-of-schedule bonding modifications to the spacecraft for Apollo 7, the first manned Apollo flight.

The following individuals were mentioned in the February 9, 1968, *Skywriter*: T.H. Encinas, Manuel Ybarra, and Albert Vega were members of an 81-man team that won the Effectiveness Stage Award for their outstanding performance in the mechanical checkout of the Saturn S-II Stage at the Seal Beach facility. The following names appeared in the March 1, 1968, *Skywriter*: Mario Bejarano, Bill Sanchez, and Edward Valles were employees in the Saturn S-II Stage Assembly Welding group who were congratulated for their top workmanship in the completion of two-cylinder weld jobs for the Saturn A-II-9 flight stage.

The following Eagle Knight was featured in the *Skywriter* in an article accompanied by a photograph: "Juan Sarabia of Tooling Support, a former janitor for an airline, is one of two employees who graduated from the Watts Skill Center in Los Angeles and received Apollo mechanical assembly training to become a template maker in the Apollo program, but still has his sights set on becoming a mechanic."[1] I say kudos to Eagle Knight Juan Sarabia for his accomplishment and for keeping his eye on the prize.

Paul Garcia, a technician in the Laboratories and Test Brazing Lab, was awarded the Silver Beaver Award, Scouting's highest national award, from the Los Angeles Area Council for "Distinguished service to the boyhood, civic, and religious organizations."[2] In addition, Julian Bernal of the Tech Library was noted for being PRIDE employee of the year at the White Sands test facility in New Mexico.[3]

While I was working on the Apollo Block II Stabilization and Control System, I met many Hispanics who were working on the Apollo program, with the following being true Eagle Knights: Larry Luera, who worked on the design of the Electrical System; Ralph Gomez, who worked on the design of the Environmental Control System; Leonard Colacion, who worked on the design of Communications System; Manny Alvares, who worked on the Communications System; Jake Alarid, who worked in Quality Control/Reliability; Hank Martinez, who worked in Quality Control; Phil Padilla, who worked in Financial; Gil Garcia, who worked in Industrial Engineering; Joe Gomez, who worked in Laboratories and Test; Herb Montano, who worked in Manufacturing; Angel Aguilar, who worked in Plastics; Renaldo Reyes, who worked in Laboratories; Joe Salazar, who worked in Manufacturing; Hector Maldonado, who worked in Field Engineering at Autonetics; Frank Serna, who worked in Communications at Autonetics; and Tony Vidana, who worked in Laboratories and Test.

The following names were published in the March 22, 1968, *Skywriter*:

> Henry Lozano, Joe Gomez, and Manuel E. Alvarez were three of five company artists recognized for their art with awards at the 15th Annual Technical Illustrators Management Association Exhibit. They, in conjunction with other company artists, did all the artwork required by the Space Division. [Of the three artists named above, I personally knew Manuel E. Alvarez, and I still have a company poster that was made from one of Manny's drawings. Another Hispanic I knew at Rockwell was Al Mejia, a fellow Trojan, who worked for Labor Relations. Al ran track for USC and lettered in 1950, 1951, and 1952—very impressive because there were not many Hispanics attending USC in the 1950s, much less serving as members of the USC powerful track team.] Manuel Armenta, Andy Ruiz, Louis Sandoval are three members of the teams from Saturn S-II Stage Assembly department who were assigned to perform hydrostatic and pneumostatic pressure tests on Saturn S-II assemblies.[4]

The following names were published in the March 29, 1968, *Skywriter*: "J. A. Silva, A.R. Feliciano, M.B. Vasquez, S.H. Benavidez, Raymond Estrada, J.E. Ruiz, R.S. Salazar are members of a team that earned the number one rating in the Saturn S-II Stage Checkout Seal Beach Stage effectiveness program. All are Space Division employees except R.S. Salazar who is a NASA employee." The following names were published in the May 3, 1968, *Skywriter*: "Richard L. Torres is an industrial engineer who has been appointed staff assistant for operations of the new company subsidiary, NARTRANS located in central Los Angeles being set up for the sole purpose of hiring and training the so called 'hard core unemployed.'" NARTRANS was headed by "Robbie" Robinson, who was African American and had been supervisor of the Verdan Computer unit of the Hound Dog Cruise Missile while I was working on that program from 1960 to 1964. Frank Cuevas of Test Planning and Stage Transportation, Stage Checkout, Saturn Seal Beach, received an Excellence Award for a Technology Utilization innovation. The June 14, 1968, *Skywriter* noted that John Perez of Photographic (whom I personally know) "has been elected to head the Pico Rivera Junior Chamber of Commerce." The following names were published in the June 28, 1968, *Skywriter*: "Ray Roman, Lydia Jimenez, Louise De Leon are three Task Force Honorees. The trio of division employees received certificates of appreciation for participation in Vice-President Hubert Humphrey's Task Force Motivation. F.A. Jacinto is one of the contributors from Manufacturing and Facilities for the win of the May Cost Reduction Buck Trimmer trophy competition."

For most of the end of 1967 and up until maybe early 1969, Chuck Markley and I were busy working on purging the flammable materials from the Stabilization and Control System. The first task at hand was to determine which materials in the displays and controls were flammable. (Fortunately, most of the displays and controls material was metal, with the electrical cables and the potting material where the cables entered the device being the only materials that could be flammable.) In addition, any nonflammable replacement material had to be durable enough to withstand the abrasive wear inflicted on the electrical cables by the test personnel and the astronaut crew as they moved inside of the capsule during the test prior to a moon mission. Chuck and I would also have to investigate whether it would be possible to arrange the cables inside the Apollo capsule in a way that would reduce the abrasive abuse that the cables received.

Chuck took on the task of determining which materials were flammable. He obtained samples of the nonmetallic materials that made up the cables and the potting compound from Honeywell for the flammability testing. A company specification delineated how to determine whether a material was considered flammable under Apollo rules by conducting tests in a closed chamber with the proper mixture of gases. When ignited by some electrical means, the material had to sustain ignition for a finite time for it to be declared flammable. If it did not ignite, it was considered nonflammable, and if it ignited but was not able to sustain ignition for the specified time, it was also declared nonflammable.

Chuck wrote procedures to test each display and control electrical cable covering and potting material sample individually for flammability. It was a long, drawn-out process, and at the end of the testing Chuck determined that only the cable covering material and the cable potting material were flammable. Having completed that task, the next step was to find nonflammable candidate materials for replacement. In addition to being nonflammable and durable, the replacement material had to be capable of being molded into a cable cover and used as a potting covering material. There were also other Apollo requirements the material had to meet, too numerous to list here.

Chuck and I met almost daily to review any problems he had encountered and to track his progress. Sometimes Chuck would suggest solutions to problems that both he and I agreed were okay, or, most likely, Chuck would solve the problems independently. The replacement of all flammable material in the Apollo Command Module was considered a high-priority task by the company and NASA because without a total solution, there would be no moon mission. Each day our management and the NASA management wanted to know how fast we were converging on a solution, so I had to know day by day how close Chuck was in his quest for suitable materials. I also coordinated Chuck's work with Honeywell because they were the ones who would have to mold the material into cable coverings and mold the potting onto the devices. At the same time, I was working on the cable rerouting problem.

In the meantime, some Eagle Knights were performing great work on future Apollo spacecraft. The following names were published in the November 15, 1968, *Skywriter*: "E.R. Sagrillo, L.B. Olvera, Tony Chacon, S. Barrios, Tom Chavez, Manny Barron, Joe Espinoza, Ray Pina are members of the team that has completed a defect-free spacecraft 113 (Apollo 16) stack through manufacturing."

Chuck Markley searched the whole Los Angeles area for materials, and finally, sometime in November 1968, he found some candidates. Through multiple tests, Chuck determined which of the candidate materials were nonflammable. The best candidate was a material called Fluorel, which smelled like cinnamon, was black in color and was supplied to Chuck by a company named Space-Flex (located about ten miles from Downey). However, Fluorel could not withstand much abrasive abuse or continuous bending of the cables, so now that problem had to be addressed. After days of work,

Chuck developed a solution to Fluorel's low tolerance of abrasive abuse, which was to put a Beta cloth booty over the potted section of the Translation and Rotation Hand Controllers and cover the whole length of their cables with a Beta cloth sheath. The solution also included placing some physical attachment point for the cables to anchor them down to restrict or minimize any excessive bending. Then there was the potting at the point where the cable entered the Translation and Rotation Hand Controllers, which needed a Fluorel booty to cover them. This step required Chuck to find a machine shop that could create a mold for the Fluorel booty and then see how that worked out. Once a mold for the Fluorel Booty had been developed, it was back to Space-Flex to have them produce the required product. Chuck made many trips to Space-Flex, too numerous to count, but my documented records show that I accompanied Chuck to Space-Flex five times in the month of December 1968. Chuck was relentless in his work and the best mechanical engineer with whom I had the privilege of working.

As happens many times, the solutions often bring with them new problems, and Chuck's solution was no exception. Beta cloth is made of fiberglass and was developed exclusively for use in outer space, but it is difficult to work with, especially when it must be sewn together (as it was in the case of the cable sheaths). One day, as I was on the phone coordinating with Honeywell, someone there suggested a solution to the Beta cloth sewing problem. It turned out that there was a small company in Minneapolis that made canvas sails for sailboats that was occasionally hired to create new sails and repair damaged sails for boats owned by some Honeywell employees. I suggested that Honeywell should investigate the possibility of having that company sew the Beta cloth booties to cover the potting and cable Fluorel coverings. However, I told Honeywell that we would have to get the sail company on the Apollo approved list if they were to do the work. All companies that did work for the Apollo program had to be on the approved list after passing the required criteria laid down by the NASA Quality Control. I told Honeywell to go ahead and contact the sail company to investigate the possibility of them as a possible solution. Honeywell's last comment to me before hanging up the telephone was that the sail company was quite small, with maybe two or three employees; I replied that the size of the company did not matter if Honeywell could get them on the Apollo approved list.

As mentioned previously, I was responsible for finding a new arrangement for the Rotation Hand Controller cables. While Chuck was busy working the problem of flammable materials, I was spending an enormous amount of time in the Apollo Wooden Mockup. The Apollo Mockup in Building 1 in Downey was used by engineering personnel to perform many of the engineering investigations and by the Apollo astronauts to familiarize themselves with the small quarters of the Apollo Command Module. I would visit the Apollo Mockup almost daily because invariably there were astronauts working in it, and I had to wait until there was only one astronaut inside instead of three so that I could take measurements and investigate possible alternate cable routes. The astronauts in the mockup with me while I was doing my work were Tom Stafford, John Young, Walt Cunningham, Al Warden, Rusty Schweickart, Dave Scott, and Jack Swigert. I would take a flexible piece of material the length of the Rotation Hand Controller cable into the mockup and lay it in different places and then watch the astronauts inside the mockup to see and record where they moved in the capsule and how much contact they had with the mock cable. I made paper and pencil sketches on large sheets of paper for all the promising new candidate cable routes.

When I had selected several new candidate cable routes in the Apollo Wooden Mockup and had the routes well documented, I would go to Building 290, where the real Apollo Command Modules were being tested. I would enter any one of the capsules that were available—Apollo 7, 8, 9, 10 or 11—and run the flexible piece of material along the new cable route and make notes or changes to the paper sketches on selected routes to install attachments for the cables. This was the final check to ensure that the selected route was right and that the cable length was long enough for the new path.

In solving the cable rerouting problem, I entered the Apollo Command Modules many times. Entering a flyable Apollo capsule was not an easy task. First, the Apollo capsules undergoing tests in Building 290 were all housed inside a "clean room" (a sealed section of the building that was environmentally controlled to a low degree level of contamination and a specific temperature, where human traffic was strictly controlled). To enter the "clean room," it was first necessary to be sanitized by getting my shoes brushed by a machine and getting all the dust removed from my clothes. Then I had to suit up by putting on a plastic head cover, donning on a pair of rayon or nylon coveralls, and covering my shoes with nylon or rayon booties; if you had a beard (as I did), it had to be covered by a white dust mask. (Suiting up in this way was required so that one's body did not contaminate the capsule with any foreign human particles.) After going through the sanitizing procedure, I entered the "clean room" through an air lock. As I passed the air lock, I got the feeling of the room's immensity. It was one large, huge building, with the ceiling about forty or fifty feet high. It was such an awesome and beautiful sight as you approached the Command Modules while they were being worked on inside the "clean room" in final preparation for delivery to NASA at the Kennedy Space Center, where they underwent final testing and countdown for a trip into space and possibly a voyage to the moon. From the outside view of the Command Module, its teardrop shape did not seem so big, but when you entered the module, you could start to appreciate the beauty of the machine, and it seemed to be large enough to accommodate three astronauts in slightly cramped quarters on their way to the moon and back. To me now, it all seems like a beautiful dream, but it is fact and part of history.

After entering the "clean room," one walked to the appropriate Command Module and stood in line because there were technicians and quality control personnel waiting to enter the capsule. Entry was restricted to only a few people at a time by a guard standing behind a chain barrier surrounding the Command Module. In addition, anyone entering the "clean room" was required to have a special pass badge that had to be visible to the guard authorizing one to enter the Apollo capsules. When your turn finally came to enter the Command Module, you had to sign the entry logbook, empty your pockets and, in the logbook, list all the equipment and materials you were taking into the Command Module. When you were finished with your work inside the Command Module, as you exited and returned to where the guard was located, you had to account for each item that you listed when you entered the capsule. Such meticulous record-keeping was necessary to avoid accidentally leaving foreign material in the Command Module that could float in zero gravity space during a mission and harm the crew or the Command Module.

Besides the material flammability issue, there were other problems facing the Apollo program that had to be solved before the program could even consider launching Apollo 7. The Guidance and Control Department was aware of some of the problems, but other challenges appeared later, out of nowhere, and generated a sense of urgency and created panic as the new problems appeared on dates that were closer and closer to the launch date for Apollo 7.

There was a NASA and North American Rockwell plan in place for achieving a moon landing to meet President Kennedy's deadline: As was mentioned before, initially, there was no plan to do dynamic space flight testing of boosters, Apollo electronic systems, heat shields, space flight techniques, and other Apollo structures. So, in 1964, the four unmanned Apollo capsules were added. The plan in 1968, after the flight of AFRMs 009, 011, 017 and 020 had qualified the heat shield, moon-rated boosters, electronic systems, and space flight techniques, was to fly manned Apollo moon-rated capsules. To fly to the moon, NASA would launch a rocket (composed of the Saturn V, the Saturn II, the Saturn IVB) with the payload necessary for the moon mission. The launch of the rocket would take place at the Kennedy Space Center, at which time the Saturn V and the Saturn II would place the Saturn IVB (with the Service Module, the Lunar Module and the Apollo capsule with the three-man crew attached to it) into Earth orbit. At this point, the Saturn IVB would fire the engine to take it and its payload out of Earth orbit and propel it toward the moon. When the assembly was on its way to the moon, the Saturn IVB engine would be cut off and subsequently separated from its payload and allowed to fall toward Earth and burn up. When the Saturn IVB separated and fell away, it would expose the Lunar Module that was attached to the Service Module. The next step would be to detach the Lunar Module from the Service Module so that it was behind the Service Module and then maneuver the capsule to attach the capsule apex to the Lunar Module in preparation for a lunar landing. The Apollo capsule, the Lunar Module and the Service Module would now be headed for the moon. As the Apollo capsule and the attachments approached the moon, the Service Module propulsion engine would fire to slow the assembly down so that it would go into moon elliptical orbit. The Service Module propulsion engine would fire one more time to place the assembly into moon circular orbit. From circular orbit, two crewmen would transfer to the Lunar Module, fire the lunar landing engine, and land on the moon. When the astronauts' mission on the moon was done, they would board the Lunar Module, fire the second section engine (using the first section as a platform), and rendezvous with the Apollo capsule that was left in moon orbit, being tended by the third astronaut. Once the Lunar Module was attached to the Apollo capsule, the two astronauts would transfer back into the Apollo capsule. When the Apollo crew was ready to go home, they would fire the Service Module engine while the capsule was behind the moon to gain moon escape velocity and head for Earth. On the way home, the Apollo crew would separate from the Lunar Module, placing it into solar orbit. The crew would then shed the Service Module just prior to Earth reentry.

Despite AFRMs 009, 011, 017 and 020 having qualified all the important elements for the lunar flights, sending an Apollo capsule directly to the moon was still an extremely risky operation, so the plan was to rehearse the moon landing path by first launching Apollo 7 into Earth orbit, thus completing the first element of the moon landing journey. Next would come launching Apollo 8 into Earth orbit, followed by sending Apollo 8 to orbit the moon and returning it home safely—the second element of the moon landing journey. Apollo 9 would then be launched into Earth orbit along with a Lunar Module and duplicate the docking of the Lunar Module with the Apollo capsule, transferring two astronauts into the Lunar Module and having them separate from the Apollo capsule, followed by the astronauts firing the Lunar Module engine to place it into a lower Earth orbit; from the lower Earth orbit, the Lunar Module crew would fire the Lunar Module engine toward the higher orbit to rendezvous and redock with the Apollo capsule, just as it would be done on the moon landing. The rehearsal of the final element of the moon landing would be accomplished by Apollo 10, which would be launched to moon orbit with the Lunar Module attached and, while in moon orbit, repeat what Apollo 9 had accomplished, heading down to moon orbit, within 10 miles of the lunar surface, and then returning and redocking with the Apollo capsule before heading home. Apollo 11 would be the official moon landing mission, repeating what Apollo 7, 8, 9 and 10 had already been through. However, the Apollo 11 mission was still full of risks because a manned spaceship landing on the moon had never been attempted.

I did not see the Apollo plan for landing on the moon by rehearsing each step written anywhere, but it was not hard to put the plan together based on the way assignments were given. As I, in retrospect, saw how we went about actually going to the moon, I realized it had all been previously planned and how clever it had been. I, of course, knew that we were going to the moon using the Lunar Orbit Rendezvous method.

Despite all the work Chuck and I were doing, and the long hours we were spending on our tasks, we were not sure we would be able to solve all the flammability material problems in time to support Apollo 7, the first Apollo manned mission (scheduled for the fourth quarter of 1968). It was already the middle of 1968, and neither Chuck nor I had a solution to the cable rerouting problem or confirmation of who was going to sew the Beta cloth covers.

19

Apollo Command Module Problems and More Eagle Knights

March 22, 1968, was the week for the Apollo 8 (S/C 103) Crew Compartment Stowage Review. This review took place in Building 1 in Downey, California, in the Apollo Command Module wooden mockup, which was set up with TV cameras and monitors so that the review teams could monitor the progress of the stowing process as it was being accomplished by William Anders, Michael Collins, and Frank Borman, the three Apollo 8 participant astronauts. In the small compartment of the capsule that was to serve as living quarters for three astronauts on a journey to the moon and back to Earth, everything they would need for the trip would have to be stowed in its proper place, as well as all the resulting garbage and human waste. The arrangement of the stowage compartments in the Command Module was a tremendous engineering achievement, and a successful moon trip would require extensive stowage preplanning and proper execution by the astronauts.[1]

The Apollo program required that equipment that was to be installed in the Apollo capsule had to receive a "fit check." The fit check was to be performed when the very first of such items was received in Downey to ensure that everything connected properly, so there would be no problems later. This process required that the Block II Rotation Hand Controllers receive a fit check of their dovetail attachments to the astronaut foldable couches and that the Block II Translation Hand Controller receive a fit check to its dovetail attachment at the appropriate point on the control panels. An internal letter dated November 8, 1967, from Bryan M. Hyde, the assistant designated subsystem project manager for the astronaut foldable couches, instructed J.J. Davis to immediately start an effort to produce the necessary gauges to interchangeably match the dovetail fitting for the Rotation and Translation Hand Controllers in their respective installation positions and that the engineering team coordinate with Davis' unit to determine the "go/no go" criteria for use of these inspection gauges.

Sometime in late 1967, a set of astronaut couches for installation into an Apollo capsule was delivered to North American Rockwell in Downey by Weber Aircraft. The Block II Stabilization and Control System Rotation Hand Controller was designed to be mounted on the armrest of two separate astronaut couches: the Spacecraft Commander's and the pilot's couch. It was also, as mentioned before, a requirement to check a Rotation Hand Controller with a fit-check tool to ensure that it connected properly with the couches. In late December 1967, Carl Conrad left a written note on my desk that documented a telephone call Carl had received from a person named Knowles, alerting Carl that the Rotation Hand Controllers were built improperly and would not connect with the dovetail on the couch armrest. Carl wanted me to resolve the problem. I telephoned Knowles and asked whether he could provide me with the name of the person on the Apollo capsule in Building 290 who could describe the problem. The names given to me by Knowles were John Hammond and Clint Bird.

Bryan Hyde called a meeting in his office on January 1, 1968, to discuss the dovetail no-fit problem. My notes and the meeting's minutes indicate that there were attendees from Apollo Design, Apollo Manufacturing, and Bryan Hyde's team. During the meeting, we agreed that a group of engineers would be assigned to check couches and Rotation Hand Controllers in house using the fit-check tool. I immediately went to Building 290, where the Apollo capsules and spare systems were located, and found two Rotation Hand Controllers that were being held there to be installed in an Apollo Command Module. I returned to Building 6 to my desk and began to make plans to check the two Rotation Hand Controllers with the fit-check tool. I soon found that there were two spacecraft that exhibited the Rotation Hand Controller dovetail no-fit problem: Spacecraft 2TV-1 (a special spacecraft undergoing thermal-vacuum tests) and Spacecraft 105 (undergoing vibration and acoustic tests). Carl Conrad had left the plant and had left me as acting supervisor on January 5, 1968.

The problem being considered was on Spacecraft 2TV-1, which was documented in disposition record (DR) #A46581, dated January 16 (we did not learn about the Rotation Hand Controller no-fit problem on Spacecraft 105 until after February 18, 1968, when an engineering order was issued that modified the astronaut couch dovetail attachment point to fit the Rotation Hand Controller). The DR on 2TV-1 was written by an employee named Hernandez (obviously an Eagle Knight) and states that when the Rotation Hand Controllers were installed on the Weber astronaut couches, the controllers did not fit. It also stated that Jim De La Rosa (written as "Dellarosa") of extension 3631, 2, 3 would support the solution to this problem.

On January 17, 1968, I went to Building 290 with Frank Foss to check one of the Rotation Hand Controllers against the fit-check tool, and it did not fit. On January 19, I checked the 2TV-1 astronaut couches' Rotation Hand Controllers dovetail connectors, and they both engaged to the fit-check tool, one better than the other. From these two tests, Tool Engineering maintained that the fit-check tool was okay and that it was the Rotation Hand Controllers dovetails that were in error. I knew that was not true because Carl Conrad's design unit had provided the couch engineers with original Rotation Hand Controller Honeywell drawings so they could build a master tool or jig to manufacture the dovetail attachment and the fit-check tool. Weber couch engineers were supposed to build their dovetail attachments to fit the Rotation Hand Controllers, not the other way around. However, the engineers' minds were made up, and for me to convince them otherwise, they had to be presented with enough information to convince them of their error. After I spent several hours reviewing all the data that could be gathered from the fit-check area, I had enough evidence to prove that the problem was not with the Rotation Hand Controller but with the astronaut couches. Someone at Weber had made an error in making the drawing for the mold; consequently, the Rotation Hand Controller did not fit on the couch armrest, nor did the fit-check tool fit the Rotation Hand Controllers, because it was made from the same erroneous master mold. Weber had screwed up, and I had to make them see it.

On January 19, engineers from Tool Engineering (accompanied by me) returned to Building 290 to recheck one of the Rotation Hand Controllers. Even with the data that

I had presented to them, Tool Engineering maintained that the fit-check tool was okay. The group performing the fit check (including me) immediately went into the high bay area of Building 290, where the Apollo capsules were being tested, and checked 2TV-1's Rotation Hand Controller and couches against the fit-check tool. We took the best couch and fit-check tool and checked the Rotation Hand Controller against the tool. The good-fit Rotation Hand Controller showed a no-fit result identical to the Rotation Hand Controller that had just been rechecked with the fit-check tool in Building 290. At this point, Tool Engineering agreed that the fit-check tool was in error. They took an action item to fix the tool.

In a later meeting with the astronaut couch engineers, after I gave a presentation with the same data and story that had been given to Tool Engineering on why it was the fit-check tool and astronaut couch dovetail connector that were in error, they agreed that the problem was not with the Rotation Hand Controller. The fit-check tool was in the process of being modified and the couches were returned to the supplier, and it was back to the drawing board for Weber, which required a re-machining of the dovetail connection on the couch armrest. Even after the Rotation Hand Controller dovetail no-fit problem had been resolved, there were people still calling me and Carl Conrad's design unit regarding the issue.

Later in the month, Carl Conrad was notified by a person named Jim Parker that the Rotation Hand Controller fit-check tool had been fixed and would be ready to be checked for accuracy on January 29. Early that day, I contacted Larry Hogan and informed him that the group was ready to check the Rotation Hand Controllers on Spacecraft 101 (Apollo 7), but Larry said to wait until Friday, February 2. On Thursday, February 1, I talked to Larry to schedule a 1:00 p.m. fit check of the Apollo 7 Rotation Hand Controllers; this time Larry told me to wait until 2:00 p.m. because the person who wrote the paperwork for the checkers to enter the spacecraft was in a meeting and would not be out until early afternoon. Larry told me he would call back as soon as arrangements were made. I called someone in Apollo Manufacturing to tell him that I would call him back to let him know the time of the fit check. Later Larry called me and told me to meet him in Building 290 at 3:00 p.m. I, along with the manufacturing individuals whom I had called to attend the fit check, met Larry Hogan at 3:00 and went through the long drill to enter the clean room and had the guard process us to enter Apollo 7. Once inside, we measured the Rotation Hand Controllers and the Translation Hand Controllers with the fit-check tool. In the end, they all passed. Later that day, in a telephone conversation with Clint Bird, I learned that Clint had gone to Weber, where three or four astronaut couches already built had not passed the new fit-check tool review. My comment to Clint was that the Rotation Hand Controller Honeywell drawing number 989503 controlled the dovetail to a tighter tolerance than Weber controlled the couch dovetail connector on the armrest. This could be one source of the no-fit problem.

It was now time to devise a plan for controlling the fit-check tool and the Rotation Hand Controller dovetail connector on the astronaut couch armrest. On February 6, 1968, a meeting on this subject was held, attended by Tom Fisher and Clint Bird from the Foldable Couch Unit; Skyler Call from Manufacturing Engineering; Steve Scarborough, Rotation Hand Controller Design Engineer from Honeywell; Lavern Johnson, Displays and Controls Project Engineer from Honeywell; and me. I had communicated the controller no-fit problem with Lavern Johnson as the situation was evolving and had invited

Lavern and Steve from Honeywell to attend the problem resolution meeting. The total problem was already defined, and everyone agreed that the fit-check tool was in error and that the foldable couch dovetail connector had to be controlled to agree with the Rotation Hand Controller dovetail. Six days after this meeting, E.S. Holt of Apollo Tool Engineering issued an internal letter that summarized the plan: a master Honeywell Rotation Hand Controller housing that included the dovetail connector would be provided to Apollo Tool Engineering by me so that a master dovetail fit-check tool could be built; a drill jig to locate the pawl hole for locking the Rotation Hand Controller on the astronaut couch armrest would be provided to Weber by North American Rockwell. Two internal letters issued by Bryan Hyde, one on February 22 and the second on May 1, 1968, which detailed the dimensions and tolerances for the dovetail connector on the astronaut foldable couch, were the last two actions taken that sealed the solution to this problem. All those involved were relieved that the problem was finally settled because NASA was now eight months away from launching the first manned Apollo spacecraft, Apollo 7, and the program could ill afford to jeopardize the schedule.

Aside from work problems to solve, there were other activities going on in Carl's unit that kept everyone from losing their minds due to the stress. In early 1968, the engineer who sat behind me in Carl Conrad's unit was Tak Shimizu. My conference table, located behind me, separated our desks, and behind his desk Tak had his own conference table (conference tables were possessed only by lead engineers). Tak's responsibilities included Apollo 13, and, invariably, astronaut Jack Swigert would come to meet with Tak to discuss the Apollo 13 system. Jack had been a test pilot at the Space Division before he was selected to be an astronaut, so many employees at the division knew him personally, and he later flew with James Lovell and Fred Haise as Command Module pilot on the ill-fated Apollo 13 mission (Jack Swigert was played by Kevin Bacon in the *Apollo 13* movie).

In my conversations with Tak, he would jokingly complain about the way things were going. I owned a small booklet composed of rectangular sheets measuring one and a half inches by two inches. In the center of each sheet was drawn a tiny square and above it the words "COMPLAINT FORM. Please write your complaint in the space below. Write legibly." So, one day, thinking to pull a fast one on Tak, I tore off one sheet from the booklet and handed it to him. Tak looked at the small sheet, read it and then proceeded to write something. My immediate thought was "The guy is really going to write something on it." Moments later, Tak handed the sheet back to me, and, to my real surprise, inside the square Tak had drawn a very tiny hand giving me the finger. After I looked at the small sheet that Tak had returned to me, I burst out laughing because I thought, "How clever; he turned the tables on me." As I looked up at Tak, we both ended up laughing.

Several months after the small sheet incident, Tak was let go by the company. He still communicated with some of his colleagues, and one learned that Tak had made a career change. He had gone to work for Otis, the building elevator company. On his first day at work, Tak was assigned to a team that would go out to building sites to repair elevators that had malfunctioned. On the team's first assignment of the day, Tak was told that the type of problem the team was going to solve that day generally took up to half a day to fix. From the work order, Tak read a description of the malfunction, and then in the team's files he found a set of schematic drawings of the Otis elevator system. Using the schematics, he made a quick analysis of the problem and identified several components that could have failed to work properly, thus causing the malfunction described in the work order. When Tak informed the team of his findings, they agreed with his logic. In a matter of half an hour, the bad component was located and replaced. Tak suggested they go get some coffee and take a long break. For the next few workdays, he continued to solve problems faster than normal, leading the team to suspect that Tak was a company-planted spy. It took Tak a bit of talking to convince the team that he was an employee just like they were.

There were many more problems with the Apollo displays and controls in 1968 than have been discussed here. The discussion of the Rotation Hand Controller no-fit problem in detail is provided so the reader can get a feel for the difficulty that was experienced by engineers when solving problems and the resources that had to be brought into play to solve various challenges on the Apollo program.

COMPLAINT FORM

PLEASE WRITE YOUR COMPLAINT IN SPACE BELOW. WRITE LEGIBLY.

PD-413

COMPLAINT FORM

PLEASE WRITE YOUR COMPLAINT IN SPACE BELOW. WRITE LEGIBLY.

PD-413

The top image is the blank complaint form Jim De La Rosa gave Tak Shimizu. Below is the filled-out form Tak returned to him (author's collection).

The launch of AFRM 020 occurred on April 4, 1968; this was the final qualification of the Apollo Command Module and the Saturn V rocket for manned Apollo missions. The Mission Control Programmer and the capsule performed their mission flawlessly even though two of the five engines of the second-stage booster (the Saturn II that was Bob Antletz's stove pipe) shut down prematurely. While in orbit around the Earth, the Service Propulsion System engine was fired to send the capsule 22,225 kilometers away from Earth and later reenter at a speed of 36,025 kilometers per hour. Because of the second-stage booster engines shutting down prematurely during the launch and the third-stage S-IVB booster failing to restart its engines, the planned reentry speed of 40,000 kilometers per hour could not be achieved. The failure to achieve the planned reentry speed, however, was not the fault of the Mission Control Programmer or of the Apollo capsule. Rather, the problem was caused by broken engine fuel lines due to pogo vibrations of the launching rocket (similar to the up-and-down motions a child experiences when jumping on a pogo stick). After this flight, the fuel lines were replaced with a different design that did not break during vibration, verified by extensive testing of the Saturn II booster. As important as this mission was to the Apollo program, it did not receive much media coverage, possibly because Martin Luther King, Jr., was assassinated on this same date, and just a few days before, President Lyndon Johnson had announced that he would not seek another term—two important events that dominated the news cycle for several days. However, despite receiving comparatively little attention, the AFRM 020 mission was nonetheless a very great and important milestone in achieving a moon landing by 1969.[2]

Once AFRM 020 had completed its mission, it had to be picked up from the Pacific Ocean (near Hawaii) and returned home to Downey, California. The following Eagle Knight's name was published in the April 12, 1968, *Skywriter*: "M.M. Sigala is a member of a 12-man Division Landing Safing Team who flew to Hawaii to pick up the spacecraft 020 command module after it was brought back after its successful 10-hour flight by the U.S.S Okinawa prime recovery ship." The main responsibility of the team was to check and make an initial evaluation of the Command Module's structure and systems, in addition to deactivating the various systems.

Spacecraft 020 was the last of the unmanned Apollo capsules that flew with the Mission Control Programmer that my team had helped design. From this point on, all flights of the Apollo capsules were going to be manned flights, and the integrity of the structure was crucial to astronaut survival.

In the meantime, other Eagle Knights were working on future Apollo capsules. The following names were published in the May 10, 1968, *Skywriter*: "Ambrose Garcia, J.B. Aragon, L.O. Telles of Apollo Structures are employees and members of the team that just completed the closeout welding on the command module inner crew compartment of Apollo Spacecraft 113 (Apollo 16) to record their fourth error-free job in a row." Ambrose Garcia was an unstoppable Eagle Knight.

After a complete analysis and evaluation of the AFRM 020, NASA concluded that the AFRM 020 mission was a success and decided that the next Apollo flight after AFRM 020 would be a manned mission using Spacecraft 101, known as Apollo 7.[3]

With the AFRM 020 successful mission and the Apollo 7 mission following on its heels, spirits at the Space Division and at NASA were running high. To keep the adrenalin pumping, the Space Division and NASA decided that there should be a baseball game between the Tigers of the Space Division and a NASA astronaut team, held at Cerritos College baseball field in Cerritos, California, to take place on Monday, June 3, 1968. The NASA team consisted of Jim McDivitt, Dave Scott, Dick Gordon, Al Worden, Russell Schweickart, Stuart Russa, Al Bean, Ed Mitchell, and Fred Haise, augmented by Hal Taylor, Jerry Lowe, Dave McBride, and Jake Smith. The Rockwell team members were Tiger player manager Joe Cuzzupoli (also shortstop); assistant manager for Spacecraft Apollo 8, Zeke Lenn, serving as catcher; Jim Phillips, pitcher; Norm Brady, first base; John Hammons, second base; John Healy, third base; Clyde Thomas, left field; John Fialko, center field; John Hill, right field. The Tigers fielded an entirely new team against the astronauts in the second half of the game.[4] My family attended the baseball game and afterward met and talked to as many astronauts as possible and got autographs from all those they talked to that evening. The following headline appeared in the company newspaper: "Spacecraft Tigers nip Astronauts in 'Spaceball' game of the age, 6 to 5."[5]

The good work of Eagle Knights continued, as reported in the company newspaper. The following names appeared in the June 14, 1968, issue:

> Jose Martinez, Ray Medina, Paul Javier, Eli De Leon, Carlos Ortiz, Ray Soto, Cruz Lerma are employees of Central manufacturing's Plastics and Advanced Projects Department and members of teams that have been lauded for completing two more Apollo error-free jobs on

NORTH AMERICAN ROCKWELL PRESENTS---
APOLLO 15 ASTRONAUTS VS. SPACE DIVISION---
"OUT OF THIS WORLD" BASEBALL GAME-------

ASTRONAUTS	SPACE DIVISION	
	JERSEY NO.	
Joe Allen	1.	Morris Wilkie
Vance Brand	2.	Eddie Kikuchi
Charles Conrad	3.	Joe Cuzzupoli
Dick Gordon	4.	John Healey
Karl Henize	5.	John Hammons
James Irwin	6.	John Fialko
Robert Parker	7.	Art Jarman
Dave McBride	8.	Ron Della Valle
Jim McDivett	9.	Gary Hunt
Harrison Schmitt	10.	George Freye
Rusty Schweickart	11.	Dave Weeks
Dave Scott	12.	Charlie Hoh
Al Worden	13.	Al Schmuck
Hal Taylor	14.	Mike Lee
Terry Neal	15.	Frank Doucette
Joe Garino (Manager)	16.	Bill Eastburn
	17.	Jim Phillips
	18.	Ted James
	19.	Julio Feliciano
	20.	Norm Brady
	21.	Pat Foley

Events Coordinated By:

James Phillips and Eddie Kikuchi

The actual team rosters from the astronauts' baseball game on June 3, 1968 (author's collection).

NORTH AMERICAN ROCKWELL PRESENTS---
ASTRONAUTS vs SPACE DIV.
"OUT OF THIS WORLD" BASEBALL GAME
JUNE 10th AT QUIGLEY FIELD, CITY OF COMMERCE
Pre-Game Festivities – 6:00 p.m.
ADMISSION BY TICKET ONLY

The author managed to get ten astronauts' autographs after the game. Can you match the autographs to the roster? (author's collection).

> Apollo manufacturing. Fernando Garcia is a member of the team from Apollo Bonded Structures that produced error-free umbilical housing cover for Spacecraft 106 (Apollo 10). The umbilical housing cover protects the large cables of bundles of wires that electronically connect sections of the Apollo rocket to one another. The bundles of electrical wires carry electrical signals from one section of the rocket to another for the rocket's operation and are called umbilical cables. The umbilical cables are critical to the operation of the Apollo Mission and must be protected from becoming damaged by debris from rocket sections separation. The cables are comparable to the umbilical cord that connects a baby to its mother.

The following names were published in the August 9, 1968, *Skywriter*: "Danny Ariaz, Del Ariaz, Chico Pacheco, Joe Salazar, Rich Naranjo, Conrad Terrazas, Raul Ponce, Mike Chavez, Joe Garcia, all employees of Apollo Manufacturing were praised for completing a major piece of work on spacecraft 110 (Apollo 14) in eight weeks instead of the normal ten weeks previously required for the same piece of work." Not only were these Eagle Knights good, but they were better than the standard.

The day after that tremendous baseball game, Honeywell informed me by telephone that they had finally gotten approval to have the small sail company in Minneapolis added to the Apollo approved list, and by late November 1968 Chuck Markley had

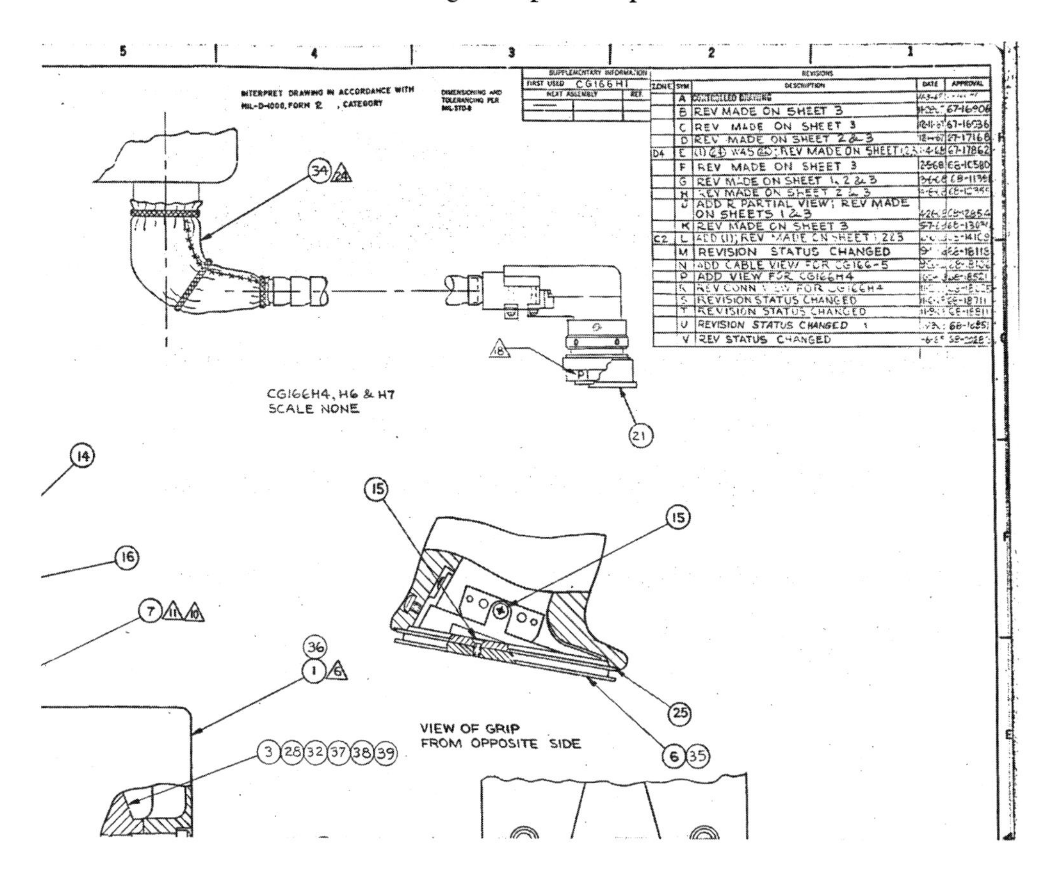

This picture is part of an engineering drawing used to build and assemble the Block II Stabilization and Control System Rotation Hand Controller (RHC). For that reason, disregard the block of notes in the top right-hand corner as well as the numbers in circles and triangles except for numbers 24 and 34 in the upper left-hand corner. The arrow from number 34 points to the elbow that was used to egress the cable that connected the RHC to the Apollo Capsule. Following the Apollo 1 fire, all flammable materials had to be purged from the Block II Apollo Capsules. The original material used for potting the elbow on the RHC had to be replaced with a nonflammable material named Fluorel. However, Fluorel turned out to be soft and easily damaged, so it had to be covered with a nonflammable protective material, known as Beta cloth. In this diagram, the elbow illustrates the Beta cloth booty (stitched by a small boat sail company in Minneapolis) over the Fluorel potting exit of the RHC connector cable (Courtesy of Honeywell Aerospace).

resolved the problems in replacing all flammable material in the displays and controls of the Block II Stabilization and Control System. Chuck found a machine shop to make the molds for the Fluorel booties for the Translation and Rotation Hand Controllers, but the solution was not timely enough to support Apollo 7 or Apollo 8, the first two manned Apollo flights. I coordinated all the work with the people at Honeywell, and they were the ones who implemented all the Fluorel and Beta cloth changes to the Rotation and Translation Hand Controllers in time to support Apollo 9.

As reasoned by NASA, it was not necessary for the Block II Stabilization and Control System displays and controls to implement the changes that purged flammable materials from Apollo 7 and 8. The NASA change control board convened on August 9, 1968, and one of the decisions the board made during this meeting was to retain the

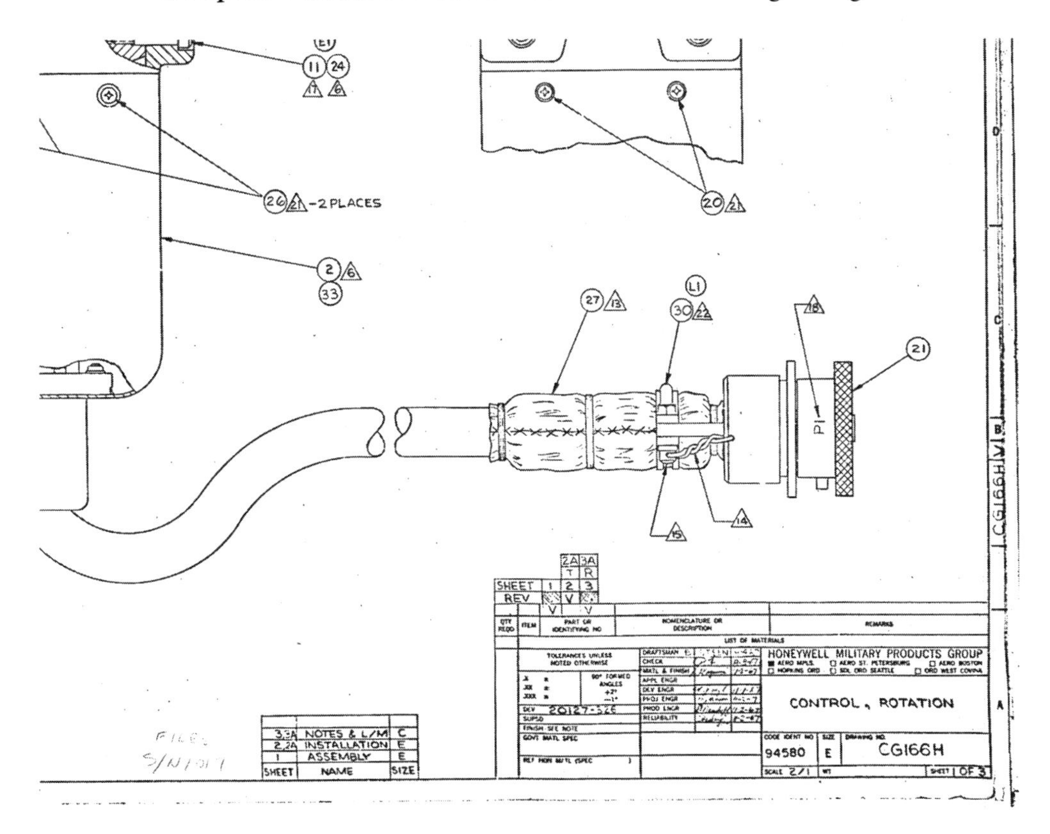

This picture is part of an engineering drawing used to build and assemble the Block II Stabilization and Control System Rotation Hand Controller (RHC). Disregard the blocks of notes at the bottom of the picture as well as the numbers in circles and triangles except for numbers 13 and 27 in the center. The arrow from number 27 points to the RHC cable, which was over six feet long. As with the RHC elbow, the sheath that covered the RHC cable was replaced with Fluorel, which had to be covered with Beta cloth, as shown in this image (Courtesy of Honeywell Aerospace).

present Rotation Hand Controller cable covering design for the Apollo 7 Command Module–Service Module. It was also decided that if the Rotation Hand Controller cables ever became damaged, they would not be returned to Honeywell for repair. Instead, a field repair on the cable covering would be made. The instructions to the Space Division from NASA were that because a field repair of Rotation Hand Controller cables had never been considered, the Space Division was to instruct Honeywell on developing a repair technique and assembling repair kits for at least four separate repairs. On August 27, 1968, North American Rockwell Space Division sent Honeywell formal authorization to implement the NASA instructions. This communication is all documented in an internal company letter dated August 27, 1968, to the contract department signed by Carl Conrad.

The following names were published in the August 16, 1968, *Skywriter*:

> Manuel Alvarez of Quality and Reliability Assurance; Jim De La Rosa, Ann Hernandez, Tony Vidana, and Rick Padilla of Apollo Engineering; Jake Alarid and Henry Martinez of Apollo Reliability; Gil Garcia, Downey Facilities and Industrial Engineering; Joe Gomez, Laboratories and Test; Ernie Fuerte, Saturn S-II Quality and Reliability Assurance, and Mary

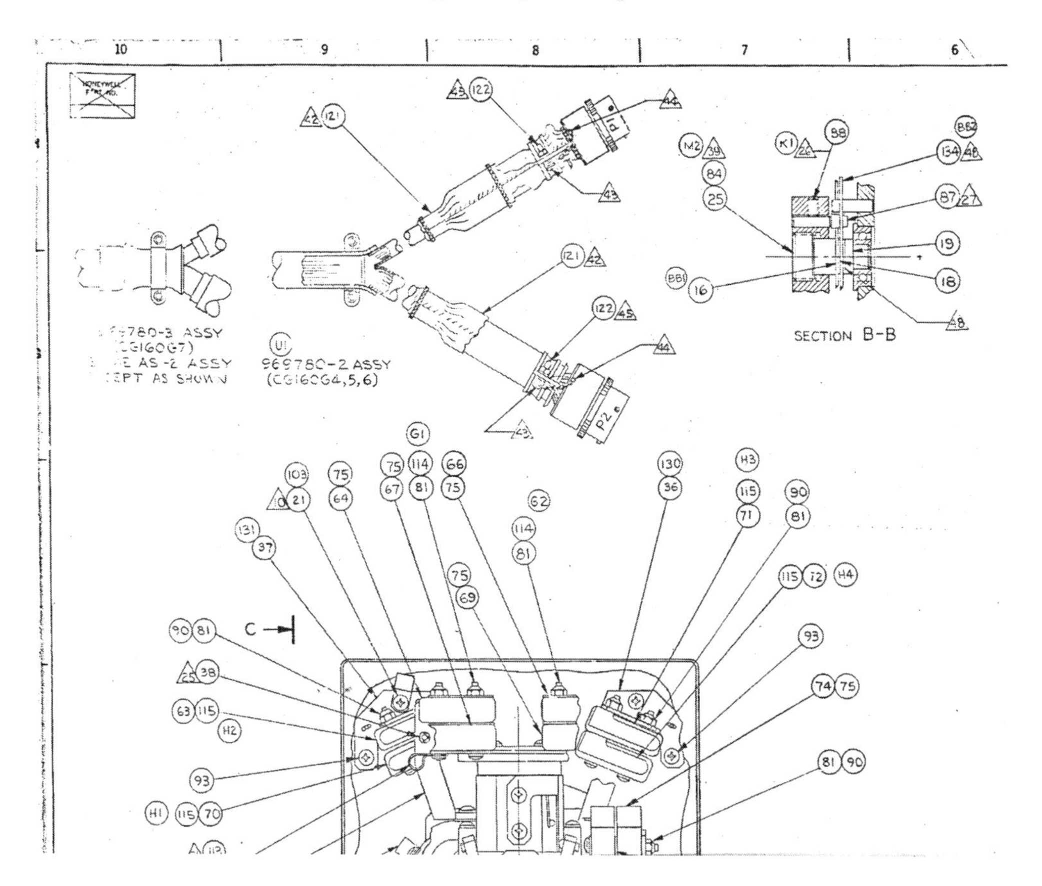

This picture is part of an engineering drawing used to build and assemble the Block II Stabilization and Control System Translation Hand Controller (THC). Disregard all the numbers in circles and triangles. The arrow at the bottom right points to one of the two THC cables, which were over six feet long and connected the device to the Apollo Capsule. Since the original material that was used for the sheath covering the cables was flammable, it also had to be replaced with Fluorel. The diagram shows the Beta cloth THC cable cover, used to protect the Fluorel (Courtesy of Honeywell Aerospace).

> Saldana and Phil Padilla of Apollo Manufacturing, all of Space Division and from Autonetics are Frank Serna of Electro Sensors Systems division; Fred Rodriguez, Navigation Systems and Tony Gaitan of Data Systems division are members of an organization, Youth Incentive Through Motivation (YITM) formed to address the school high dropout rate of Hispanic students.

The probable reason why NASA was willing to allow Apollo 7 to fly with the original cable covering design is that NASA felt the changes in the internal atmosphere in the capsule cabin (from 100 percent oxygen to a mixture of 60 percent oxygen and 40 percent nitrogen), in combination with the removal of many flammable materials from the cabin, minimization of ignition sources, and addition of a fire extinguisher along with a new hatch for astronaut quick egress, had already reduced the risk of another fire within the capsule.

As each Apollo Command Module was delivered by manufacturing to Building 290 in Downey, where it would undergo systems installation and test, each module (starting with Apollo 7) was assigned to a team of managers, project engineers, and test

Apollo Hispanic engineers. Standing row (left to right): Manny Alvares, Frank Serna, Joe Salazar, Angel Aguilar, and Leonard Colacion. Seated row (left to right): Joe Gomez, Jim De La Rosa, Jake Alarid, Hector Maldonado, Ralph Gomez, Renaldo Reyes, Hank Martinez, Phil Padilla, and Herb Montano (Courtesy of the Boeing Company).

personnel that would push them through Building 290 for delivery to NASA at Kennedy Space Center in preparation for launch. The July 12, 1968, issue of *Skywriter* featured A.B. Kehlet as the assistant program manager responsible for getting the Apollo 10 and Apollo 11 Command Modules through the Downey facility and delivered to NASA on schedule. Al Kehlet would later become chief engineer on the Space Shuttle program, working for Buzz Hello, who in the 1970s became the vice president and program manager of the Space Shuttle program.

The following names were published in the July 12, 1968, *Skywriter*: "Ray Sena was the leading man in the annual division Tennis Tournament. Ray won the trophy in the men's open singles, won the men's double title with teammate George Shull, and the mixed doubles trophy with his wife Pat." I personally knew Eagle Knight Ray Sena. In addition, R.L. Morales of Saturn S-II was awarded a twenty-year service pin.

In the beginning of 1968, Chuck Markley and I were so busy that no trips to Honeywell were necessary until later. It turned out that all Honeywell design and test and document approval tasks were ones that I could approve by telephone. The design changes made to the Block II Stabilization and Control System were of two types: Class One changes affected form, fit and function of the devices and needed NASA and division management approval, whereas Class Two design changes did not. All the system design changes that I dealt with were of the Class Two type, which required only the company's engineering approval. Because the displays and controls were 90 percent mechanical devices, most of the changes that Honeywell requested were made for the purpose of facilitating the manufacture of the devices or to solve some problem encountered during testing. The design and test document changes were usually single changes and were easy to approve over the telephone. It was only when completely new design or test documents or total document revisions had to be approved that a trip to Honeywell to review and give signature approval for the complete document was required.

My first trip to Honeywell in 1968 took place on May 22, which was nine months after my return from Minneapolis. The Block II Stabilization and Control System had long ago completed its Environmental Qualification Test, but there were significant design changes made to the system afterward that it required additional qualification testing to verify these design changes under space environments. The testing was defined in a package named "Delta Qualification Test." It was for this Delta Qualification Test that my trip to Honeywell was scheduled. The tests had been satisfactorily completed, and Orville Littleton (NASA subsystem manager from the Mission Control Center in Houston) and I were meeting at Honeywell to review the results of the tests with Honeywell engineers; if all was as it should be, Orville and I were to give our signature approval for its completion to certify that the system as designed was qualified for manned space flight. Everything in the review with Honeywell went smoothly, and by May 24 Orville and I were heading home.

Besides my assignment to visit Honeywell and take care of business, there was also the ongoing need to address problems at the home plant with Chuck Markley and other situations that needed attention, so it was not like I was coming back to paradise. To fill my plate a little bit more, Carl Conrad was gone from the plant from July 23 through July 25 and assigned me as acting supervisor while he was gone. The second trip to Honeywell was on September 3, 1968, for a design review of the Rotation and Translation Hand Controllers with Orville Littleton and the Honeywell engineers. The purpose of this meeting was also to supplement existing information on the capability of the controllers to meet the Apollo mission requirements, which was required because of the impending Apollo 7 flight the following month. With the completion of all tasks, it was back to Downey for me on September 6.

After I returned to Downey from Honeywell, the company newspaper published more updates about Eagle Knights. The following names appeared in the September 13, 1968, *Skywriter*: "Edward Quinones, Orland Marin are two Inspectors from S-II Quality and Reliability Assurance who were members of the Saturn S-II Stage Insulation Bond and Test team that has compiled a perfect 18-month no defect record at Seal Beach in the application of cork insulation to the Saturn S-II booster stage." (How about that? Perfect work completed on Bob Antletz's stove pipe, the second rocket section on the Apollo moon rocket that was attached to the Saturn V booster.) And then, on September 20, 1968: "A.L. Salcido is a member of the Apollo Bonded Structures team that accomplished perfect workmanship on its eighth Apollo command module inner crew compartment. The inner crew compartment will serve as the astronaut crew habitat in the Apollo capsule." The following note was published in the September 20, 1968, *Skywriter*: "Ann Hernandez is a member of the Dixie Belles, division women's softball champions."

The September 27, 1968, issue of *Skywriter* included the following notification: "Andrew Ruiz, Juan Villarreal are two of 119 veteran employees having 25 and more years of service with the company who were honored at the annual division dinner on Thursday night, September 26." To put the 25 years that these two Eagle Knights had served this company into perspective, one must recall that they started working for the company in 1943. In September 1943, World War II had been raging on for one year and nine months. At that time, I was thirteen years old and in the seventh grade. In the intervening 25 years, I completed grammar school, high school, one semester at UCLA, and four years of service in the Air Force, after which I worked for Hughes Aircraft Company six years, followed by eight years at Rockwell; I had received my engineering degree from USC and been married for 14 years, and my four children had already been born. In that same time, these men had been building careers at Rockwell; if that is not impressive, I do not know what is.

October 13, 1968, was the date of my third trip to Honeywell, and, being autumn, it was already beginning to get a little cold in Minneapolis. This trip was intended to resolve problems associated with the design and fabrication of the Rotation and Translation Hand Controllers. During my stay in Minneapolis, the Apollo 7 crew was orbiting the Earth, having been launched into space on October 11 and scheduled to splash down on October 22. My return home was on October 18, several days before Apollo 7's splashdown. (There will be more to say about Apollo 7 in a later chapter.)

My next trip to Honeywell was on November 7; I was heading out to resolve some Rotation Hand Controller problems that Honeywell was having. Carl Conrad was gone from the plant for another three days from November 18 through November 20, and he again appointed me acting supervisor, which meant that in addition to my regular duties, there were staff meetings to attend and other supervisory duties to perform. This just made life more interesting. It was also in late 1968 that Carl's unit was reorganized.

The following names were published in the November 8, 1968, *Skywriter*: "Martin Gutierrez of Apollo Engineering, Orlando Marin, Edward Quinones, both of S-II Quality and Reliability Assurance, Carlos G. Ramos of Central Material were awarded Snoopy Pins [the astronauts' personal symbol of excellence on the Apollo/Saturn program] Wednesday, November 6, by astronauts Wally Schirra and Walt Cunningham."

Minneapolis was already in its winter season when I made my last trip to Honeywell for 1968. This trip was on December 4, with my return to Downey scheduled for December 6. On this visit, we would determine the degree of repairs that had to be made to a Rotation Hand Controller that had been dropped and received high-impact "G" forces. The question was whether the Rotation Hand Controller had survived the high-impact forces with minimum damage and could be repaired or whether the controller had been damaged beyond repair and should be scrapped. To verify the condition of the Rotation Hand Controller, it had to be electronically tested and disassembled and all parts inspected visually and x-rayed—all of which had to be witnessed by me to help Honeywell decide to either salvage or scrap the Rotation Hand Controller. There was also the problem of resolving configuration engineering problems with the delivery of the first Fluorel nonflammable Rotation and Translation Hand Controllers for the Apollo 9 Command–Service Module.

20

Manned Apollo First Flight and More Eagle Knights

The first few manned Apollo flights, including the first moon landing mission, were the most challenging and the most exciting because of all the unknowns and the anticipation of each mission's outcome. Remember that at this time no manned spacecraft had ever been into outer space outside the influence of Earth's gravity, let alone to the moon and back. The Apollo program did not just put three astronauts in an Apollo capsule and send them to the moon. The journey to the moon was planned and rehearsed over a period of seven years. As described in earlier chapters, the plan started with boilerplates that were put through various environmental and structure strength tests. Along with the boilerplates, there were four Block I Apollo capsules (AFRMs 009, 011, 017 and 020) that were flown unmanned by the Control Programmer and the Mission Control Programmer to qualify many of the systems, the heat shield, and numerous flight techniques over a period of three years. The manned Apollo capsule flights, starting with Apollo 7 through Apollo 10, flew in stages to verify every aspect of the journey to the moon short of landing on the lunar surface. These preparatory journeys were not easy, so read the remaining chapters to learn about the behind-the-scenes activities that will tell the story of how Apollo 11 was able to land on the moon to win the space race for the United States.

The lead engineers who were responsible for each system that was installed in the Apollo capsule and other system specialists were required to support the manned missions in real time. There was an area in the Guidance and Control Lab in one of the buildings in Downey, California, that had been set up with strip chart Sanborn Recorders to record data from the Command and Service Modules as the mission was taking place. Those engineers selected to support the Apollo manned flights in real time were to man the stations in shifts on a twenty-four-hour basis from launch time to splashdown of the Apollo Command Module on its return to Earth. For the Guidance and Control Department, the assignments generally were that Paul Garcia and I would split a twenty-four-hour shift; Paul was on for twelve hours and I was there for the next twelve hours, being joined by specialists Jack Jensen, George Cortes and Mert Stiles. The recorded data was transmitted via the telemetry system from the Command Module to receiving stations on the ground so that engineers could monitor the performance of their respective system during the mission. We engineers in the Guidance and Control Lab were in communication with the NASA engineers in Houston in case NASA needed our technical assistance. The support of Apollo missions for the Space Division in this fashion started with Apollo 7 and continued to the last Apollo flight.

To me, all the manned Apollo launches up to Apollo 11, the first landing on the moon, appeared to have occurred in rather rapid succession because all the dedicated work on the part of the Apollo program at North American Rockwell Space Division had paid off, and the division was ready with hardware, software, procedures, and engineering staff for the manned flights. The technical staffs at NASA and the industrial companies that had brought the Apollo program to fruition had worked hard to hone their skills and were never more ready to send three men to the moon.

In 1964, when I started working on the AFRM 009 project, no one had ever soft landed a space vehicle on the moon, but in 1966 through 1968 Hughes Aircraft Company, through a contract from the Jet Propulsion Laboratories in Pasadena, California, financed by NASA, had made five successful soft landings (out of seven total attempts of landing on the moon) with the Surveyor unmanned space vehicle, which was a critical requirement for landing men safely on the moon. Those soft landings of Surveyors provided a wealth of data about the moon's surface and certified the moon journey and terminal guidance for moon soft landing, proving that a manned vehicle landing on the moon was possible. In addition, the Surveyors and the Ranger program returned valuable information on possible landing sites that was needed for a safe Apollo moon landing.[1]

At the time of the Apollo missions, the cost of putting spacecraft or satellite payloads into orbit was about $1,000 a pound. The goal of the space program was to bring this cost down to $100 a pound, decreasing the cost by an order of magnitude. To accomplish this goal, the workers in the space program were demanding that the designers and manufacturers of engines create more powerful and smaller engines, that lighter and stronger materials be developed by the materials industry, that more powerful fuels be developed, and that the electronics industry provide smaller, more powerful and faster electronic components. In October 1969, at a Space Shuttle symposium held in Washington, George Mueller presented these opening remarks: "The goal we have set for ourselves is the reduction of the present cost of operating in space from the current figure of $1000 a pound for a payload delivered in orbit by the Saturn V down to a level of somewhere between $20 and $50 a pound."[2]

Replacement of analog electronic circuits with digital electronic circuits saved the space program thousands of pounds of hardware and copper wiring. Imagine so much copper electrical wire being replaced by two wires, similar to a telephone cable used for the internet in which digital data of zeros and ones could be transmitted instead of thousands of individual analog electrical signals being transmitted via a thousand individual copper wires. The smaller and faster digital electronic components and small microprocessors, along with the various application software programs developed for the space program, were what eventually led to the development of the handheld calculator and paved the way for the PC and cell phones. The handheld calculator allowed scientists and engineers, in the 1970s, to discard the old-fashioned slide rule. With more powerful electronic microprocessors, it was possible to replace hardware with a more powerful microprocessor and software to put a payload into orbit that was lighter and that had greater data-processing capabilities.

Only aerospace workers like the ones named in this book and those who worked for other aerospace companies can appreciate how far technology has come since those days. In 1958, when I was working for Hughes Aircraft Company as a test equipment design engineer and the Hughes radar engineers were designing the digital computer for the Hughes Radar MA-1 Fire Control System (to be installed in the newest interceptor-fighter airplane at the time), the digital computer was hardwired logic, whereas modern computers are driven by software logic. The Hughes computer was designed with miniature vacuum tubes (as opposed to today's computers, which are designed with solid-state semiconductor logic in electronic chips) and was as big as a desk, with a magnetic drum memory that had been verified at Lincoln Laboratories at MIT. By today's standards, that early computer was tremendously large, with a tiny memory that contained numbers used for calculations, and small in processing capacity. The 100 hours of digital computer instruction that I received back in 1958 at Hughes included digital logic, digital arithmetic with fixed-point and floating-point arithmetic, Boolean algebra, truth tables and binary numbers. The digital computer class also included many hours of instruction on how the Hughes computer in the MA-1 System worked. Everything I learned in class at that time is still valid today, but since then the biggest changes have been in computer size and capacity, software programs and the proliferation of more electronic gadgets for people to play with. All this development has been possible because of the miniaturization of electronic components and the software that came out of the space program, which allowed engineers to pack more computing power into smaller electronics and create more powerful software, which is what is driving today's economy.

Here is an example of something that was developed during the space program for those of you who know very little about it or consider the space program a waste of money: In 1967, at the peak of the space program, while working for the Jet Propulsion Laboratories in Pasadena, Dr. Andrew Viterbi proposed a dynamic programming algorithm to be applied to noisy digital data codes communication links. This software algorithm separates the information from the noise in digital data communication links, such as cell phones and digital TV, so you can keep the true digital information and throw away the noise. Prior to this invention, digital communication by phones was very noisy. The cell phone uses a digital communication link and is manufactured with miniature electronics that resulted from components developed during the space program, but without the Viterbi Algorithm incorporated into the cell phone, the conversation would be so noisy and garbled that zero cell phones would be sold. Thanks to Dr. Viterbi, the cell phone economy is surging and there are millions of people out there with cell phones that contain a Viterbi Algorithm chip. Other electronic digital products in which the Viterbi Algorithm chip is used are modems and digital TVs. As a result of this development, Andrew Viterbi has become a billionaire, and he and his wife made a multimillion-dollar donation to the USC school of engineering, my alma mater, where Dr. Viterbi received his doctorate, with the simple request that the engineering school be named the Viterbi School of Engineering; how is that for driving today's economy from the wasted money spent on the space program?[3]

In addition, the PC and laptop were not invented by Steven Jobs or IBM, nor were software operating systems like Microsoft invented by Bill Gates. In the 1960s and 1970s, computers were large mainframe computers owned by large companies. In those companies, for a computer programmer to get his program into workable software, he had to convert his program code into Hollerith Punch Cards and submit his card stack for overnight processing by the large mainframe computer; his results would be ready for review the next morning. During those same years, the aerospace industry was reducing the size of digital computers for their specialized use. A few examples are the miniature version of the digital computer that weighed around sixty pounds, such as the one used on the Hound Dog cruise missile, the Minute Man ICBM and other ICBMs, and the Space Shuttle as well as the Apollo Command Module Computer (CMC). IBM and Steve Jobs merely repackaged those types of digital computers for commercial sale and personal use, and thus the PC was born. As mentioned earlier, up until the time that Dr. Ramon Alonso designed the CMC, small computers did not have keyboards, but Dr. Alonso put a keyboard in his CMC to give the astronauts the ability to enter data into the computer, thereby giving IBM and Steve Jobs a paradigm for their PC. The electronic components used by Jobs and IBM were those solid-state components that had been miniaturized for the space program. If it had not been for the space program, we would probably not have the PC, the internet, the smartphone, the laptop, the iPad, digital TV, communication satellites, weather satellites, or GPS satellites.

Getting back to 1968, once the Rotation Hand Controller no-fit problem had been solved and the Apollo 7 Weber astronaut foldable couches were fixed, it was time to move the Apollo 7 Command Module forward even though Chuck Markley and I could not complete the Fluorel hand control nonflammable changes in time to be incorporated into the first manned Apollo Block II capsule to go into space. It was now the week of May 6, 1968, and Command Module was in the process of undergoing the crew compartment fit and function check with astronauts participating before being prepared for shipment. On Saturday, May 11, 1968, the Command Module got a taste of its destiny in Building 290 in Downey when it was tested for the first time by the Apollo 7 crew under a simulated flight environment. Astronauts Walt Cunningham, Walter Schirra (flight commander), and Don Eisele completed the crew compartment fit and function check to historically put on the books another completed Apollo 7 milestone.[4]

The Apollo 7 Command Module completed all the required testing in Building 290 and was delivered to NASA on Wednesday, May 29, 1968.[5] Following its delivery to NASA, the Command Module was housed in the vertical assembly building (commonly referred to as the VAB) at the Kennedy Space Center to undergo more testing and stacking with the Service Module and the required boosters for the historic manned Apollo maiden flight. On the week of August 2, 1968, the Command Module for Apollo 7 was ready for the trip to the launch pad from the VAB via the Crawler (a massive steel flat bed with tractor wheels that had the capacity to carry a 12-million-pound rocket from the vertical assembly building to the launch pad), and on Friday, August 9, 1968, it was transferred to the launch pad on a platform of the Crawler for further testing and an October 11, 1968, launch date, as announced by General Samuel Phillips of NASA to usher in the manned Apollo flights.[6]

In accordance with the required protocol, Honeywell sent a letter to North American Rockwell confirming that the Stabilization and Control System delivered for Apollo 7 was certified for flight. In return, a letter dated September 5, 1968, was sent to H. Matisoff of Contract Administration by Carl Conrad, approved by Bill Fouts, instructing Matisoff to tell Honeywell that Rockwell approved of the Apollo 7 certification that Honeywell had made and to advise Honeywell that the Guidance and Control Engineering Department agreed that the Apollo 7 Stabilization and Control System was ready for flight.

Apollo 7 was launched on October 11, 1968, with splashdown occurring on October 22. The crew made 163 orbits around Earth, paving the way for the first leg for Apollo 11's moon journey, and was a complete success. As mentioned before, the lead engineers at NA Rockwell Downey were required to support the Apollo missions in the laboratory where recording equipment and communications were hooked up for that purpose; that's where I spent most of my time during that mission, except for the few days that I had to visit Honeywell in Minneapolis. Other engineers who shared the twenty-four-hour vigil with me were Jack Jensen, George Cortes, Mert Stiles and Paul Garcia.

A 1995 photograph of the vertical assembly building at Kennedy Space Center, where Apollo Rockets were assembled in a vertical position and tested prior to taking them (via the Crawler) to the launch pad (author's collection).

The Apollo 7 Command Module was back in Downey, California, for postflight analyses on November 1, 1968. The ablative heat shield on the Apollo 7 capsule was black and charred, and the burned streaks next to the windows were there just like on AFRM 009, but this time there had been three astronauts in the capsule who could look out the window and see the flames as they came up the side of the capsule. On Wednesday, November 6, 1968, we went with members of our organization to greet the astronauts of the historic Apollo 7 flight with all employees on the grounds of the Space and Information Systems Division. Only Walt Cunningham and Wally Schirra showed up; Don Eisele, the third member of the crew, was unable to attend due to illness. They were greeted by more than 8,000 division employees on their return to Downey.

While the Apollo 7 Command Module was back in Downey, it was checked out by a team of employees that included some Eagle Knights. The following names were published in the November 1, 1968, *Skywriter*:

> Lino Salazar Space Division technician, Hector Rodriguez NASA Project Engineer are members of the team that is performing the post-recovery checkout of the Apollo 7 command module following its 4.6-million-mile perfect mission in space. Apollo 7 was the first manned Apollo capsule to ever take flight into Earth orbit. The first leg of the journey to the moon of Apollo 11 is to launch the first moon lander Apollo capsule into an Earth orbit like Apollo 7 so, Apollo 7 was a rehearsal of the first leg of Apollo 11 to its moon journey. The next Apollo capsule, Apollo 8, will be a trip to the moon with moon orbits followed by a return to Earth which will be a rehearsal of the second leg of Apollo 11's moon landing journey.

As important as the Apollo 7 mission was, it was equally important that the proper postflight analyses be conducted. The Apollo program had to know what had gone right so it could be repeated on the following mission, Apollo 8. More important, it was

The author and a closer view of the Crawler that shows how large it is in a photograph taken at Kennedy Space Center in 1995 (author's collection).

A closer view of Complex 39, including the launch pad. In the foreground is the author with a camera bag on his shoulder at Kennedy Space Center in 1995 (author's collection).

imperative that the program be aware of what went wrong so it would not be repeated on subsequent missions. It was also economically important to discover what degradation the onboard systems suffered while exposed to the environment of flying into space so that improvements could be made and also so that a decision could be made regarding whether to use some of the systems' equipment on future flights or maybe even use some parts in Apollo capsules that were not designed to fly in outer space, thereby saving money.

For all the reasons listed above, Ed P. Smith, manager of Apollo CSM Project Engineering, issued a seventeen-page letter (#695-200-020-68-111) dated September 24, 1968, with the title "CSM 101 (Apollo 7) Post Flight Testing Plan of Action," which defined the plan for all the Apollo Command Module postflight hardware testing. The testing was anticipated to start approximately on October 28 and to be completed by December 23. On October 31, 1968, an engineering order, originating in the Guidance and Control Department, was released to comply with Ed Smith's letter. The order was written by Bernice Johnston to authorize the Material Department to direct Honeywell to perform postflight tests on the Apollo 7 hand controls for anomaly verification and to measure any degradation that might have occurred. The three devices from Apollo 7 that were submitted for testing were Rotation Hand Controllers serial numbers 1043 and 1044 and Translation Hand Controller serial number 1020. The engineering order also defined the tests to be conducted and their sequence.

Honeywell released an interoffice correspondence dated November 12, 1968, that provided detailed procedural instructions for the postflight testing of the two Apollo 7 Spacecraft Rotation Hand Controllers; it was signed by six Honeywell employees, as well as H. Matisoff and myself from NA Rockwell. On November 15, Honeywell released a second letter that provided detailed procedural instructions for the postflight testing of the Apollo 7 spacecraft Translation Hand Controller. November 15, 1968, was also a historic day at NA Rockwell. That was the day when it was announced that Bobbie Johnson was promoted to manager, the highest post for a woman in engineering thus far. This advance was the beginning of change resulting from the 1960s activity regarding equal opportunity and affirmative action.

Of the three hand controllers in the Apollo Command Module, only the Rotation Hand Controller serial number 1043 was reported to have exhibited an anomaly during the Apollo 7 flight. This anomaly was that when a minus pitch minimum impulse maneuver was initially attempted to fire some reaction jets, it was successfully executed, but when a second minus pitch minimum impulse was attempted, the vehicle did not respond. The minus pitch breakout switch was suspected to have malfunctioned. The three controllers were removed from the Command Module after the flight and returned to Honeywell for test. Rotation Hand Controller serial number 1043 was to be tested and analyzed for the cause of the reported flight anomaly, and the other two controllers were to be analyzed to determine the amount of degradation suffered during the flight.

Honeywell began the postflight analysis test by subjecting Rotation Hand Controller serial number 1044 to an acceptance test and discovered that the press-to-talk switch on the grip handle was malfunctioning, so on November 20, 1968, Honeywell sent a request to North American Rockwell for permission to remove the trigger assembly from the grip of the controller for analysis of the suspected malfunction of the communication switch. I received a Honeywell message dated November 26, 1968, from Bernice

Johnston that said Honeywell had removed the trigger assembly from the grip assembly and now needed permission to remove the grip assembly for internal visual examination. The message had a stapled note attached to it to Bernice Johnston from Larry Workman in Program Administration asking Bernice to ask me to send him a speed letter to authorize Honeywell to remove the grip assembly. Larry received the requested speed letter from me, and that same day he sent Honeywell a wire authorizing them to remove the grip assembly from the Rotation Hand Controller for internal visual examination. This last authorization from me sent to Honeywell allowed them to analyze the cause of the suspected malfunction of the press-to-talk switch and to continue with the testing of the remaining two Apollo 7 controllers. By the end of the first week in December, Honeywell had completed the testing and analyses of the three Apollo 7 devices. Honeywell was kind enough to mail me a draft copy of the final analysis report so that I could write my report on the Honeywell findings to NASA.

Using the advanced copy of the Honeywell Apollo 7 controls postflight analysis report, I wrote an Engineering Summary Report with the title "CSM 101 (Apollo 7) Hand Controls Post Flight Analysis" that was signed as written by J. De La Rosa and signed as approved by Bob Epple and Larry Hogan. The report documented the results of the postflight analyses of Apollo 7 Command Module Rotation Hand Controllers serial numbers 1043 and 1044 and Translation Hand Controller serial number 1020. The tentative conclusion resulting from the Honeywell tests was as follows: Testing and analysis of Rotation Hand Controller serial number 1043 resulted in no evidence of malfunction to substantiate the astronaut-reported in-flight anomaly. The two most likely failures that could have caused the anomaly were the minus pitch cam follower or the minus pitch breakout switch. An exhaustive search of the Rotation Hand Controller serial number 1043 history revealed that there was no track record of problems or failures associated with the minus pitch breakout switch. However, the minus pitch cam follower had several failures that were attributed to a sticky cam follower caused by burrs left in the cam follower bushing-type-bearing during manufacturing re-boring operations. However, since Honeywell could not duplicate the malfunction that was reported during the flight, nor could they find evidence that the Rotation Hand Controller had malfunctioned at all, their conclusion was that the malfunction had not occurred.

When I submitted the first rough draft of my report prepared for NASA to Carl Conrad for his review, complete with Honeywell's conclusion that the malfunction reported by the astronauts during the flight had not occurred because of lack of evidence to support it, Carl bounced it back to me with a note instructing me to rewrite the conclusion. No doubt Carl did not want to give the impression that Wally Schirra's crew had screwed up. The final report that I circulated for approval and that everyone signed stated that NA Rockwell considered a minus pitch sticky cam follower the most probable cause of the in-flight anomaly based on previous failure history data. A momentary sticky breakout switch could also have caused the problem, but there was no failure history indicating this failure mode was likely to occur. To preclude the sticky cam follower from ever causing the problem again, a design change was implemented in the Rotation Hand Controller by Honeywell for the Command–Service Module for Apollo 8 and subsequent Command Modules, which replaced the bushing-type-bearing with a low cam follower-to-shaft friction ball bearing in the pitch, yaw, and roll cam followers. The redesign of the Rotation Hand Controller also included a change in the breakout switch

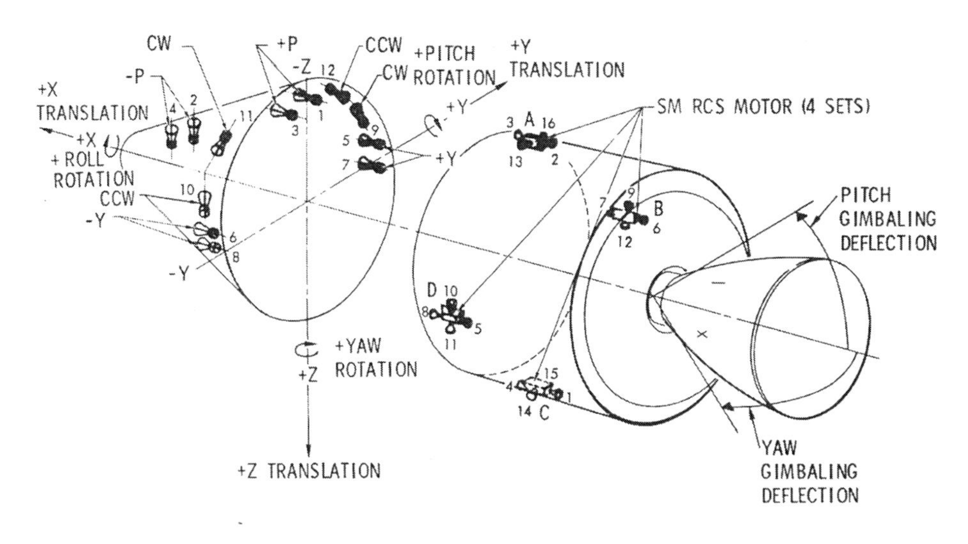

Figure 5-2. Thruster Orientation

The capsule control jets were fired only on Earth reentry. The Service Module control jets and the gimballed engine were fired only during the moon trip and Earth return trip (Courtesy of the Boeing Company).

from the Texas Instruments to the Honeywell-type switch, which test data verified as an improvement. The redesign of the two most probable causes for the anomaly put this failure mode in the closed category with no further action required, giving the redesigned Rotation Hand Controllers a clean bill of health for the Apollo 8 flight.

The second Apollo 7 Rotation Hand Controller, serial number 1044, was initially submitted by Honeywell to incoming functional testing in which the communication press-to-talk switch exhibited less than full travel when depressed. The result of the investigation was that a small mechanical component in the trigger assembly was being impeded by excessive crystal formation. Further teardown indicated that crystals existed deeper inside the device grip assembly. The salt crystals that caused the press-to-talk switch to "freeze up" were found to be sodium chloride (sea salt). It could only be concluded that the sodium chloride resulted from saltwater splashed on the Rotation Hand Controller while the Command Module was in the water with the hatch door open after splashdown. Testing qualified the Rotation Hand Controllers for a salt solution spray of 1 percent salt by weight in distilled water, which was enough to qualify the Rotation Hand Controllers for whatever salt content humidity existed in the Command Module cabin due to astronaut body perspiration. The Rotation Hand Controllers were not intended to qualify for a saltwater immersion test from sea water. This discrepancy was closed with no further action required.

Translation Hand Controller serial number 1020 was functionally tested at ambient temperature and passed all test requirements. Since visual inspection and functional tests showed no evidence of discrepancies, it was concluded that no more effort should be expended on this device. On December 20, 1968, Honeywell transmitted the postflight analysis report for Apollo 7 Rotation Hand Controllers serial numbers 1043 and 1044, dated December 12, 1968, complete with a cover letter, to C. Conrad with the same conclusions as in the report draft they had given to me.

As a check and evaluation of the performance of the Apollo 7 Block II Stabilization and Control System Attitude Reference System, which was a backup to the primary Inertial Measurement Unit, engineer Bob Zermuehlen released an internal letter dated January 9, 1969, with Supervisor Carl Conrad's signature, titled "Results of Apollo 7 Attitude Reference Comparison Test and Backup Gyro Display Coupler Alignment Test." Conclusions of the tests were that both test objectives resulted in performance well within the maximum allowable limits. The largest system drift away from a nominal value was 3.0 degrees per hour in the pitch axis (the allowable system drift per specification was approximately 10 degrees per hour). The backup Block II Stabilization and Control System alignment accuracy of approximately 0.25 degrees was likewise considerably better than expected. On January 30, 1969, Rockwell sent a wire to Honeywell instructing them to return all the Apollo 7 tested hardware to Rockwell in "as is" condition. This event concluded the Apollo 7 Rotation and Translation Hand Controllers' postflight analyses.

The backup Attitude Reference System from Apollo 7 was sent to Honeywell for analysis. Afterward, Honeywell sent North American Rockwell a Subcontractor/Supplier Document Review dated May 13, 1969, with an attached Supplement 1 Post-Flight Analysis Report Apollo 7, dated May 1, 1969. The report was on the electronic backup unit serial number 1007, written by engineer Don Reynick, approved by Chuck Moosbrugger (Apollo Engineering project engineer) and other Honeywell personnel. At this time at Honeywell, there were many relays contaminated inside the outer case with solder balls, and disassembly of Backup Unit 1007 from Apollo 7 found one relay that was contaminated with one solder ball inside the case. All x-rays of all relays were thoroughly reviewed for solder ball contamination, and all contaminated relays were sought out to remove them from devices where they had been installed. This problem was resolved prior to the flight of subsequent Apollo flights. If this issue had been allowed to persist, the solder balls could have shorted out the backup unit, meaning that the Apollo capsule would no longer have a backup Attitude Reference System; such a failure could spell a catastrophe in case the primary system failed on the way to or back from the moon.

Between the splashdown of Apollo 7 and the launch of Apollo 8, Supervisor Carl Conrad released two internal letters (one dated May 29, 1968, and the second dated July 23, 1968) stating that he left me acting supervisor while he was absent from the plant. As mentioned before, whenever Carl appointed me acting supervisor, it added further tasks to my workload, but I knew that Carl depended on me, and filling in for him gave me experience that I could put on my resume. A supervisor or management position for me at Rockwell was out of the question by this time because of company culture and my being a rabble rouser with very high company exposure during participation in activities to make changes and improve upper mobility opportunities for minority groups at NA Rockwell Space Division.

21

Prelude to a Moon Landing with More Eagle Knights Participating

After the successful launch and Earth orbital mission of Apollo 7, it was time to prepare for Apollo 8, the first real moon mission and a rehearsal for the second leg of the moon journey for Apollo 11. Up to this point, this was the most important mission because Apollo 8 was scheduled to make several orbits around the moon in December 1968. This would be the first manned Saturn V booster launch and the first shot at verifying the guidance and navigation techniques to the moon and back using Dr. Ramon Alonso's Command Module computer, as well as an opportunity to perfect the communication techniques between Houston and the crew in an Apollo capsule in an actual moon mission. It would also be the first closeup view of the moon surface by humans and the first human view of the back side of the moon.

North American Rockwell Space Division had been working hard to get the Apollo 8 Command Module ready for delivery to NASA. The Apollo 8 spacecraft Customer Acceptance Readiness Review between the NA Rockwell Space Division and NASA was held during the week of June 7, 1968.[1] This was the same type of review presentation that was discussed before, in which the NASA high-level executive board would sit up front, joined by the NA Rockwell Space Division high-level executives, with microphones in front of them. The NA Rockwell Space Division lower-level managers sat down in front of the audience with their teams ready to answer, through roving microphones, questions that came up regarding the readiness of the spacecraft for delivery to NASA. The Apollo 8 Customer Acceptance Readiness Review was successfully completed, and the Apollo 8 Command Module was shipped to Kennedy Space Center by the company earlier in the week prior to Friday, August 16, 1968.[2] The launch for Apollo 8 was planned for December 21, 1968, which was announced in the *Skywriter*; the company newspaper had a photograph of the crew and an article that said that the official launch countdown for Apollo 8 at Kennedy Space Center was scheduled to begin Monday, December 16, 1968.[3]

Apollo 8 was launched at 4:51 a.m. PST, Saturday, December 21, 1968. Astronauts Frank Borman (commander), Jim Lovell (Command Module pilot) and Bill Anders (Lunar Module pilot) were in the Apollo Command Module during the launch, sitting atop the Saturn V booster that was made up of 5.6 million parts and tons of explosive fuel.[4] On Saturday, the first day of the mission, after the crew had checked out the vehicle in Earth orbit and confirmed that it was ready for a lunar trip some two hours after launch, Michael Collins, the capsule communicator on the ground, gave the Apollo 8 crew approval from Houston for a trans-lunar injection, at which time the S-4B booster reignited its engine to boost Apollo 8 on its trip to the moon. The additional power of the S-4B engine increased the velocity of the spacecraft to overcome Earth's gravity. The Apollo 8 mission was one of the most emotional ones for the Space Division and the world at large when the astronauts read from the Bible and wished everyone a merry Christmas during one of their moon orbits on Christmas Eve. The crew also took pictures as they circled the moon.

Sunday, the second day of the Apollo 8 mission, was partially devoted to the first television transmission by the crew from space, with millions of people watching worldwide. On Monday, the astronauts completed star-Earth horizon sightings with the onboard optical alignment system and the sextant; the data from the sightings could be used on board or transmitted to the ground for Apollo 8 space navigation. Available to the astronauts aboard the spacecraft for star-Earth horizon sightings were a scanning telescope for wide-angle star location and a sextant for the actual star sighting. In addition to being transmitted to ground-based computers, the navigation information the astronauts gathered could be entered into the capsule's onboard Command Module Computer by the crew via the onboard display keyboard (known as the DSKY, pronounced "dis-key").[5]

With the DSKY, the astronauts could enter verbs and nouns. Verbs requested that some action be taken, while nouns requested data; they were entered into the DSKY by codes. For example, suppose the crew entered a make-believe verb code 22 into the DSKY, and this code was interpreted as "Fine Align IMU"; the computer would then electronically fine align the Inertial Measurement Unit used for Command Module onboard navigation. Or suppose the crew entered a make-believe noun code 41 into the DSKY, which was interpreted as "Time of Landing Site"; in response, the computer would provide the crew with the time of their landing site. The verb and noun codes that I have given as examples here are modeled from real codes in a pamphlet given to me by one of my Guidance and Navigation colleagues at work in 1969. It contains all the guidance and navigation information the Apollo crew needed on board the capsule for communicating with Ramon Alonso's computer via the DSKY during a moon mission.[6]

At the time that the astronauts had performed the star-Earth horizon sightings, it was not known whether Apollo 8 was going to circumnavigate the moon to permit the astronauts to continue a free-return-to-Earth path or to fire the Service Propulsion System engine to place the Apollo Command Module into moon elliptical orbit. After evaluating all the spacecraft information, Mission Control at the Manned Spacecraft Flight Center in Houston gave the Apollo 8 crew the go-ahead to place the spacecraft into moon elliptical orbit. On Tuesday, the fourth day of the mission, the astronauts fired the Service Propulsion System engine again to bring Apollo 8 out of the elliptical orbit and into a circular orbit around the moon at about 70 miles above the lunar surface—the closest a human had ever been to the moon. Tuesday was Christmas Eve, and Commander Frank Borman and the other members of the crew read from Genesis in a broadcast that was heard all over the world. After 10 lunar orbits, the astronauts fired the Service Propulsion System once more on the back side of the moon to take the spacecraft out of moon orbit and into a trans-Earth trajectory.

On Wednesday, Christmas Day, during the trans-Earth coast, the astronauts had turkey for one of their meals and kept referring to our planet as the "Good Earth." On Thursday, December 26, following additional horizon sightings for navigation, the astronauts performed minor mid-course corrections to ensure that they were inside the tubular corridor for Earth reentry. After certifying that the Apollo 8 spacecraft was on the proper course, the crew proceeded to transmit ground views of the Earth and its continents. Friday, December 27, was splashdown 1,000 miles south of Hawaii, being placed there by a rolling entry maneuver orchestrated automatically by Ramon Alonso's Command Module Computer. After a successful Earth reentry and landing in the Pacific Ocean at 7:51 p.m. PST, six days after the December 21 lift-off from Cape Kennedy, the Apollo 8 capsule was picked up by the aircraft carrier *Yorktown* after a journey of half a million miles. The Apollo 8 Command Module was returned to Downey, California, and was available for viewing by employees and their families in Building 47 from 6:00 to 10:00 p.m. on Friday, January 3, 1969, and from 9:00 a.m. to 5:00 p.m. on Saturday and Sunday. My family was not going to miss seeing the first Apollo Command Module that circled the moon, so we were there during that weekend.[7]

The Apollo 8 crew and their spaceship accomplished many firsts. Some are listed below, and they are unquestionably impressive:

- "First manned flight of the Saturn 5 launch vehicle.
- First manned flight of the Saturn S-2, second stage of the Saturn 5.
- Highest altitude for a manned mission.
- First manned flight to the vicinity of the moon.
- First lunar orbital flight.
- First photographs and television transmitted by men from the lunar vicinity.
- First restart of the third stage engine during a manned flight.
- Heaviest payload for a manned mission into orbit—about 280,000 pounds.
- Greatest distance mankind has ever been from earth.
- Highest speed ever attained by human beings—more than 24,000 miles per hour, and
- First direct human view of the dark side of the moon."[8]

The following is the prayer that Commander Frank Borman beamed to Earth during the third lunar orbit:

> Give us, Oh God, the vision which can see thy love in the world in spite of human failure.
> Give us the faith to trust the goodness in spite of our ignorance and weakness.
> Give us the knowledge that we may continue to pray with understanding hearts and show us what each one of us can do to set forward the coming of the day of universal peace, Amen.[9]

Some 1,200 reporters (including more than 100 reporters from foreign countries) covered the launch of Apollo 8 at Kennedy Space Center. Thanks to the availability of television to people all around the world, the launch was viewed by more persons than any previous event in human history. Several days before the Apollo 8 launch, no hotel rooms or flights into Cocoa Beach, Florida (which is adjacent to Kennedy Space Center), were available, so people slept in Melbourne (25 miles south of the launch site), or even Orlando (40 miles west), and drove north the following day to witness the great event.

Live telecasts from Apollo 8 were beamed directly by satellite to Europe and Central and South America. In addition, Italian, Japanese and Mexican newscasters broadcast the launch live from Kennedy Space Center. A telecast from Apollo 8, showing the lunar surface, brought forth the following praise from the London *Evening Standard*:

"Circling the moon are men like us—braver and more skillful than we are, but still our bone and blood. And that is a reason for added rejoicing this year."

The whole world celebrated the Apollo 8 astronauts' Christmas Eve orbits of the moon. In countries where Santa Claus is not part of the culture or where the birth of Christ is not celebrated, the Apollo 8 astronauts were toasted as the heroes of the hour. Very few humans criticized the awesome voyage. The most notable was Samuel Shenton, secretary of London's Flat Earth Society, who commented that the "public is being ballyhooed, taken for a ride." This person, with no knowledge of science and no understanding of what was happening, was present for one of the greatest achievements of the 20th century and managed to miss it as though he was not even there. Apollo 8 commander Frank Borman, who was a graduate of Tucson Magnet High School in Arizona, was honored by a parade in Tucson sometime after his return from his historic moon mission. His picture hangs at Tucson High School, along with those of other famous alumni, in celebration of his place in history; the picture was still there when I had the honor of visiting the school in 2013.[10]

The Downey facility where three sections of the Apollo 8 rocket were designed and manufactured was full of Eagle Knights. The following names were published in the January 24, 1969, *Skywriter*: Don Gallegos of Administration is one of many outstanding contributors in their win of the Apollo Program Buc Trimmer Trophy, awarded monthly to the company organization that trims the highest dollars from program costs. Al Alcantar of NASA is a member of Al Kehlet's top team responsible for managing the processing of spacecraft 107 (Apollo 11) from its beginning almost five years ago to its delivery to the NASA Wednesday January 22, 1969. Spacecraft 107 will be the first Apollo to transport a crew of three astronauts for the first man landing on the moon by the end of 1969. Al Alcantar was a NASA Resident Project Engineer. A.R. Salcido is one member of the Apollo Bonded Structures team that accomplished perfect workmanship on Apollo command module inner crew compartment.

According to the February 7, 1969, *Skywriter*, "S.I. (Jose) Jiminez was awarded a Snoopy Pin [the astronauts' personal symbol of excellence on the Apollo/Saturn Program] last week by astronaut Al Worden." (This is the same Eagle Knight mentioned in the August 25, 1967, *Skywriter* as a member of the team that went to different U.S. cities briefing people on the importance of Spacecraft 017 in the Apollo Program.) Later, the following name was published in the February 14, 1969, *Skywriter*: "Carlos Diaz was the Supervisor of the team that attained a delivery date goal of a Saturn booster stage." The February 14 *Skywriter* also announced that former Hound Dog cruise missile program chief engineer Gary Osbon, now Apollo chief engineer, had been appointed vice president of Products Planning, Commercial Products Group of North American Aviation.

There were five manned Apollo flights scheduled for 1969, and Apollo 9 was the third crewed mission and tremendously important for the Apollo program. For the first time, the launch vehicle would carry a Lunar Module. The engineers from my department supported this flight as they did the previous Apollo flights; however, no notes can be found from my support of Apollo 9. What is in my possession is the Flight Support Crib Sheet given to the personnel who were supporting the mission. The crib sheet has detailed information of every phase of the mission and even some information that was not released to the public.

The road to getting Apollo 9 ready for its mission was not an easy one and involved many employees at the Space Division, ultimately culminating in delivery of all the

Apollo 9 components to NASA. For this mission, the updated Rotation Hand Controllers and Translation Hand Controllers that Chuck Markley and I had worked on were included in the spacecraft. The January 10, 1969, *Skywriter*'s front page had a full-size picture of the thirty-six-story-tall Apollo 9 rocket when it was rolled out of the vertical assembly building at the Kennedy Space Center on the way to the launch pad via the Crawler, where it completed an extensive pre-lift-off checkout. The twelve-million-pound vehicle traveled to the pad at an average speed of half an mile per hour.[11]

Apollo 9 was an Earth orbital mission, never leaving the confines of Earth's gravity, but it would be the first time that the Apollo capsule connected with the Lunar Module in space, followed by two astronauts transferring from the Command Module into the Lunar Module in space; at that time, the Lunar Module would separate from the Apollo capsule, and the astronauts would maneuver the Lunar Module to a lower Earth orbit to later rendezvous with the Command Module, just as it would happen during the first moon landing.

The launch countdown of Apollo 9, as the February 21, 1969, *Skywriter* reported, was to begin Saturday, February 22, in preparation for the launch that would take place on February 28, 1969, at 7:00 a.m. California time. However, as was posted in the company newspaper a week later, the Apollo 9 launch was rescheduled to Monday, March 3, 1969, due to astronaut cold symptoms.[12] Despite this delay, Apollo 9 launched successfully and accomplished all its mission objectives, returning to Earth on Thursday, March 13, 1969. The Command Module splashed down in the Atlantic Ocean, with astronauts Jim McDivitt (commander), Dave Scott (Command Module pilot) and Russell Schweickart (Lunar Module pilot) on board, thus bringing the Apollo program one step closer to a moon landing mission. This vital mission was the first step in qualifying the Lunar Module, the only component of the moon landing mission that had not yet been proven with men aboard in a space environment.[13]

The list of Hispanics working on Apollo to the moon keeps growing, and more Eagle Knights were featured in the company newspaper. The following names appeared in the March 14, 1969, *Skywriter*: "Leadman J.P. Contreras, R.R. Hernandez, C.C. Rangel, Fernando Ventura, J.R. Jaramillo, Apolonio Ramirez, E.R. Valles, and leadman L.A. Garcia are members of the team from S-II Seal Beach who were commended for welding two major units of the Saturn S-II booster free of defects." (Recall that the Saturn S-II booster, Bob Antletz's stove pipe, was the second stage of the Saturn V booster stack required for an Apollo moon voyage.)

Apollo 9 was unloaded from the USS *Guadalcanal* at Norfolk, Virginia, on the first leg of the journey back to California. The Command Module was returned to Downey after its 10-day sojourn in space and was available for viewing by employees and families in Building 247 on March 21–23. The module underwent extensive post-flight tests, all of which concluded that the flight was a huge success. A special Space Division "Welcome Back" took place on Thursday, April 10, 1969, for astronauts Jim McDivitt, Dave Scott, and Russell Schweickart, who returned to the division for the first time since their milestone mission. The trio visited Seal Beach, California, and after the ceremonies there they visited Downey and were greeted by a total of ten thousand employees at the dual ceremonies at the two facilities. The visit was arranged by the astronauts.[14]

The February 28, 1969, *Skywriter* announced that a Space Shuttle contract had been awarded to the Space and Information Systems Division by NASA's Manned Spacecraft

Center to perform a six-month conceptual study to investigate the possibility of designing and building a low-cost, manned logistics (Space Shuttle) system using present-day technology. This contract was one of four issued by NASA to aerospace companies. In addition to the Manned Spacecraft Center, other project studies were being directed by NASA's Langley Research Center in Virginia and the Marshall Space Flight Center in Alabama. Known as Integral Launch and Reentry Vehicle (ILRV) studies, the contracts concerned different aspects of future Space Shuttle vehicles. Division work on the study was done by a team at the Seal Beach facility. John Sandford had overall supervision; George Fraser was study manager. All facets of division technology were involved in the performance of this work. Representatives on the team were from Structures and Design, Central Manufacturing, Science and Technology, Management Planning and Control, and Contracts and Pricing. This was an important contract because the division had to seek future work to hold on to its employees whenever the Apollo program ended. NASA's interest was in space exploration, and the Space Shuttle was a key element in future space endeavors and a step in the right direction for NASA.

Another contract award to the Space Division of NA Rockwell was announced in the April 4, 1969, *Skywriter*. The contract was with the NASA Langley Research Center for a 10-month $155,000 feasibility study to design a two-man "lunar emergency escape-to-orbit vehicle." The proposed vehicle was intended to be carried aboard the Lunar Module and could be used by the astronauts to return to the Command–Service Module in case of a Lunar Module failure on the moon's surface. The escape vehicle was being designed as a "light empty-weight, minimum complexity two-man lunar escape system utilizing unsophisticated guidance and control techniques and simplified propulsion and stabilization and control concepts."

More Hispanics who worked on the Apollo program were featured in the company newspaper. The following names were published in the April 4, 1969, *Skywriter*: "Ray Sena, and Jess Orona are guards on the Hawks Space Division basketball team and Tony Avelino is a team member of the Green Giants Space Division basketball team. The Hawks won the Space Division basketball championship against the Green Giants. Frank Vigil is one of the persons to call in Seal Beach for tickets for the Sixth Annual Reunion of the Navaho Program Test Group being held on April 18, 1969, at the Jolly Roger Inn on Katella in Anaheim." The Navaho program was a piggyback missile contract awarded to the North American Aviation Missile Division in the mid–1940s and consisted of the Navaho booster rocket and the X-10 cruise missile. The contract was cancelled in the mid–1950s in favor of intercontinental ballistic missiles. Many of the engineers I worked with during the Hound Dog cruise missile program and Apollo program had worked on the Navaho program, including Frank Vigil, who was working with the Hound Dog cruise missile test team in Florida from 1959 to 1964.

The following name appeared in the April 11, 1969, *Skywriter*: "Mike Rodriguez a Space Division Downey employee is a member of the Space Division summer bowling league." The April 18, 1969, *Skywriter* featured more names: "Fred Rodriguez from S-II Field Support Seal Beach, Leo Garcia, and J.A. Medina of S-II Quality and Reliability Assurance Seal Beach were awarded Snoopy Pins by Apollo 9 astronauts."

The following names were published in the April 25, 1969, *Skywriter*: "Charles I. Macias an employee of Space Division is father of Charles R. Macias who is one of 21 boys who won a company 1969 college scholarship. Jim De La Rosa, Joe Gomez, Manny Alvarez, Ralph Gomez, Angel Aguilar, Art Ochoa, Bill Ruiz, Caesar Sanchez of Space Division and Hector Maldonado, Frank Serna, Al Chavez are members of Youth Incentive Through Motivation, an organization that is involved in bringing programs to the local schools with the objective of bringing the Mexican American student drop-out rate down and encourage a continuation of their education." I met all of the above-mentioned Eagle Knights while I was a member of the Youth Incentive Through Motivation organization.

The following names were published in the May 2, 1969, *Skywriter*: "Robert Jimenez of industrial management and J.J. Hernandez of systems engineering with their specialties indicated are two of 29 employees who have received master-degrees under a company educational plan. Frank Cuevas, Jr., of Saturn S-II received a 30-year company service pin." It must be noted that Frank Cuevas was a true Eagle Knight who started working for North American Aviation in 1939 during the Great Depression; by 1969, he had spent nearly half a lifetime with the company. "S.I. (Jose) Jimenez of Apollo Logistics Training is one of two experts who will work with Public Relations in a news desk based at Downey, California, that will provide technical support to aid newsmen in the Western states during the Apollo 10 mission. Other company experts have been selected to provide technical support for media coverage at Kennedy Space Center and other experts were selected to provide additional technical support in a Downey remote studio to ABC, CBS, and NBC to supplement their media support for their Kennedy Space Center and New York flight coverage."

The following names were published in the June 27, 1969, *Skywriter*: Gilbert Garcia, chief of the manufacturing Operations Section of NASA-Downey, R.E. Valencia of Manufacturing Subsystem Assembly, and L.R. Sanchez were three people involved in an Apollo milestone. Garcia accepted, for NASA, the section for an Apollo Service Module that was manufactured by the Space Division team of which Valencia and Sanchez were members. I knew Gilbert Garcia of NASA personally. Gil and I met as students at East Los Angeles Community College in 1954. We were both surprised to find that Gil had served in the U.S. Marines during the Korean War with a John De La Rosa who turned out to be my older brother.

To our delight, for those of us on the Apollo program, the June 6 and 13, 1969, issues of the *Skywriter* announced that a July 16 date had been set for Apollo 11, which would put three American astronauts in space for a moon landing mission, as reported by George Low, NASA Apollo program manager.[15] This was exciting news for everyone at NA Rockwell because now there was a date to shoot for to beat the Russians to the moon.

Originally, NASA had scheduled the flight of Apollo 10 for May 17, 1969, but they rescheduled the launch date to May 18, the second day of the lunar launch window. If ever there was an important Apollo flight, Apollo 10 was it. This mission was going to do almost everything that Apollo 11 would do except land on the moon. The Apollo 10 Lunar Module, with two astronauts aboard, would approach the moon and come within approximately 50,000 feet of touching down on the surface. If this step could be done successfully, surely Apollo 11 could land on the moon in July 1969. (Remember that by this time the unmanned Surveyor had already made at least two soft landings on the moon.)

Apollo 10's Flight Readiness Test was completed in the week of April 4, 1969, at the Kennedy Space Center. The objective of this test was to check out all electronic systems to ensure that they were ready for flight. Apollo 10 was launched from Kennedy Space Center on Sunday, May 18, 1969, with Tom Stafford, John Young and Gene Cernan aboard. On Thursday, May 22, Commander Stafford and Lunar Module Pilot

Cernan entered the Lunar Module and separated from the Command-Service Module, descending until they were less than 10 miles from the lunar surface, while John Young remained in moon orbit as the Command Module pilot. The following day, May 23, the Lunar Module and the Command-Service Module rendezvous occurred 8 hours after the original separation. Later that same day, the Lunar Module was jettisoned into solar orbit. During the Apollo 10 mission, the crew provided the first color TV moon and space views for those watching on Earth. Now Apollo 10 was ready to come home, and on Saturday, May 24, 1969, the Command-Service Module fired the Service Propulsion System engine (the system that was the realm of George Cortes' expertise) for trans-Earth injection, which increased the velocity of Apollo 10 to break away from the moon's gravity and return to Earth.[16]

The Apollo 10 Command Module was on display in Building 247 in Downey, California, on Friday, June 6, 1969, from 5:00 to 9:00 p.m. and Saturday and Sunday from 9:00 to 5:00 p.m. As in the past, my family took advantage of the chance to come see a true piece of history. Like the other Apollo capsules after they returned from their space missions, the blunt end of the capsule was black and absolutely charred. The heat shield, being made of some material composed of a filled epoxy resin, bonded in fiberglass honeycomb, partially burned off because of the friction during the fiery reentry.[17] On Thursday, June 19, 1969, the Apollo 10 astronauts returned to Downey to thank a crowd of 7,000 employees. The astronaut trio was also greeted by thousands at a parade in downtown Los Angeles.

The Apollo 10 crew. Thomas P. Stafford and Eugene A. Cernan, when inside the Lunar Module (known as Snoopy), descended within 8 nautical miles of the moon's surface (author's collection).

As an example of the type of notes I took during my time supporting the Apollo flights, my records from Apollo 10 have been included in the appendix. To make the notes easier to understand, I have added explanations in italics. Be advised that there are several pages in the appendix; however, if you are a real Apollo aficionado, you will enjoy reading these notes. It was not easy taking notes during a live moon mission, and some of the statements are cryptic. Those who are interested can probably

NORTH AMERICAN ROCKWELL CORPORATION
Space Division
12214 Lakewood Boulevard
Downey, California 90241

APOLLO 10 SPACECRAFT COMMAND MODULE

The command module for the Apollo 10 mission was launched on its journey to the moon from Kennedy Space Center, Fla. Crew members were Astronauts Thomas P. Stafford, commander; John W. Young, command module pilot; and Eugene A. Cernan, lunar module pilot.

The Apollo 10 command module orbited the moon 31 times (62 hours) before heading for home.

Moon Orbit Altitude	69 statute miles
Mission Duration	192 hours (eight days)
Distance Traveled	830,726 miles (round trip)
Maximum Speed	24,759 1/2 miles per hour (36,315 ft. per. sec.)
Flight Path Reentry Angle	-6.52 degrees
Maximum Heat (during reentry)	4,200 degrees Fahrenheit (Approx)
Weight at Launch	12,265 pounds
Weight at Splashdown	10,860 pounds
Spacecraft Serial Number	SC-106
Code Name for CM	Charlie Brown
Code Name for LM	Snoopy
Significant Dates	
Begin Structural Assembly	June 24, 1966
Manufacturing Complete	July 15, 1968
Checkout Complete	Oct. 19, 1968
Shipped from Downey	Nov. 26, 1968
Launch	10:49 a.m. (PDT) Sunday, May 18, 1969
Enter Moon Orbit	1:34 p.m. Wednesday, May 21, 1969
Leave Moon Orbit	3:09 a.m. Saturday, May 24, 1969
Splashdown	9:52 a.m. Monday, May 26, 1969
Return to Downey	June 4, 1969

Mission Evaluation: "110 per cent success"--Glynn Lunney, flight director, Manned Spacecraft Center.

A description of the Apollo 10 mission (Courtesy of the Boeing Company).

obtain a copy of any of the recorded live Apollo flights, but with all the NASA jargon used during the missions, it is doubtful that they will make much sense without some interpretation (like the one I have included).

One Honeywell postflight analysis report for Apollo 10 flight equipment, written by Don Reynick and approved by Chuck Moosbrugger and others, was delivered to Supervisor Carl Conrad's design unit around the end of July. The title of the report was "Post

Flight Analysis Report for Gyro Assembly S/N 1017 and 1022 Plus Gyro Display Coupler S/N 1007 from S/C 106 (Apollo 10)," and it was dated July 16, 1969. Tests were conducted on these three devices as part of an investigation into abnormal Stabilization and Control System Attitude Reference System drift rates reported by the Apollo 10 astronauts. The conclusion was that the tests at the device and system level did not confirm the excessive drift rates.

Even though we had completed four Apollo manned flights—one Earth orbital mission, one mission that verified the docking of the Command Module with the Lunar Module, and two missions that involved journeying to the moon and back to Earth—things were not completely perfect heading into the scheduled moon mission of Apollo 11. There was a flaw in one of the moon mission critical systems that had exhibited itself in two of the Apollo flights, and, if not corrected for future Apollo missions, it had the potential to spell disaster for the spacecraft and their crews.

22

Solving One Last Major Problem Before the Moon Landing

The closer the launch date of Apollo 11 approached, the more high profile any problem became, with more pressure for its solution coming from the executive community at the Space Division. Some Block II Stabilization and Control System problems occurred in March, but they will not be detailed here. The problem discussed in this chapter, like the Rotation Hand Controller no-fit dilemma (previously discussed), exemplifies the processes the program had to go through to solve any critical issue. It is a behind-the-scenes Apollo story that has never been told, so read on for new revelations, as well as a surprise solution.

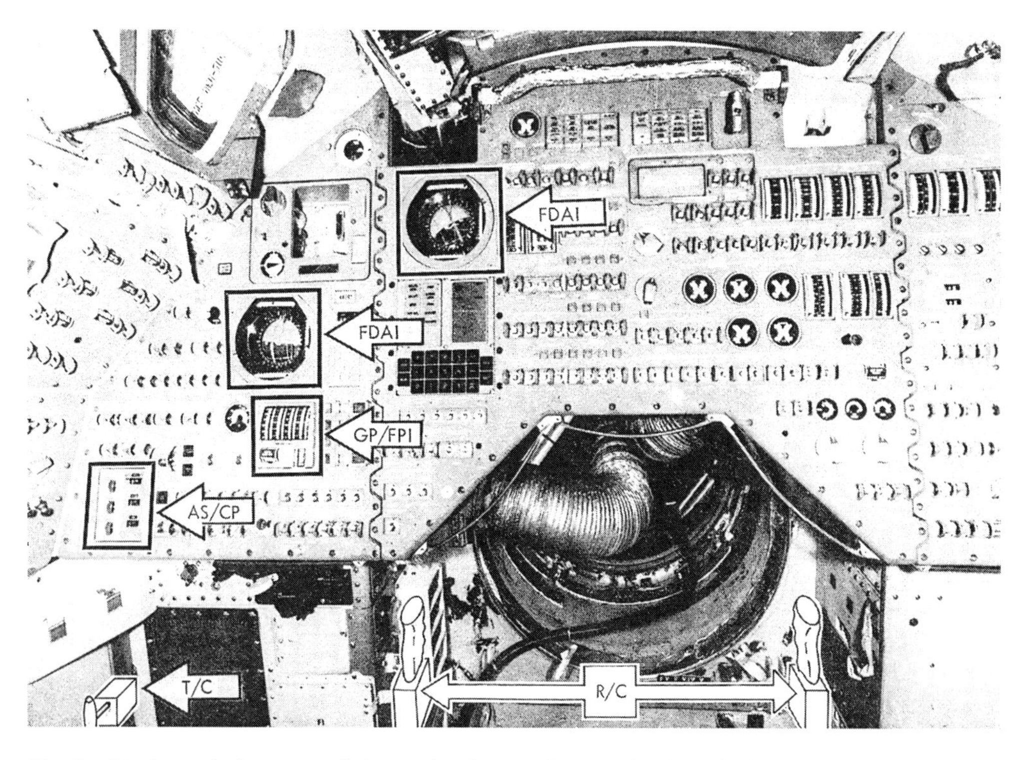

The Apollo Capsule front panel shows displays and controls identified with arrows. The circular tunnel below the main panel is the passage for the crew between the capsule and the Lunar Module. The Entry Monitor System is at the top left next to "FDAI" (Courtesy of the Boeing Company).

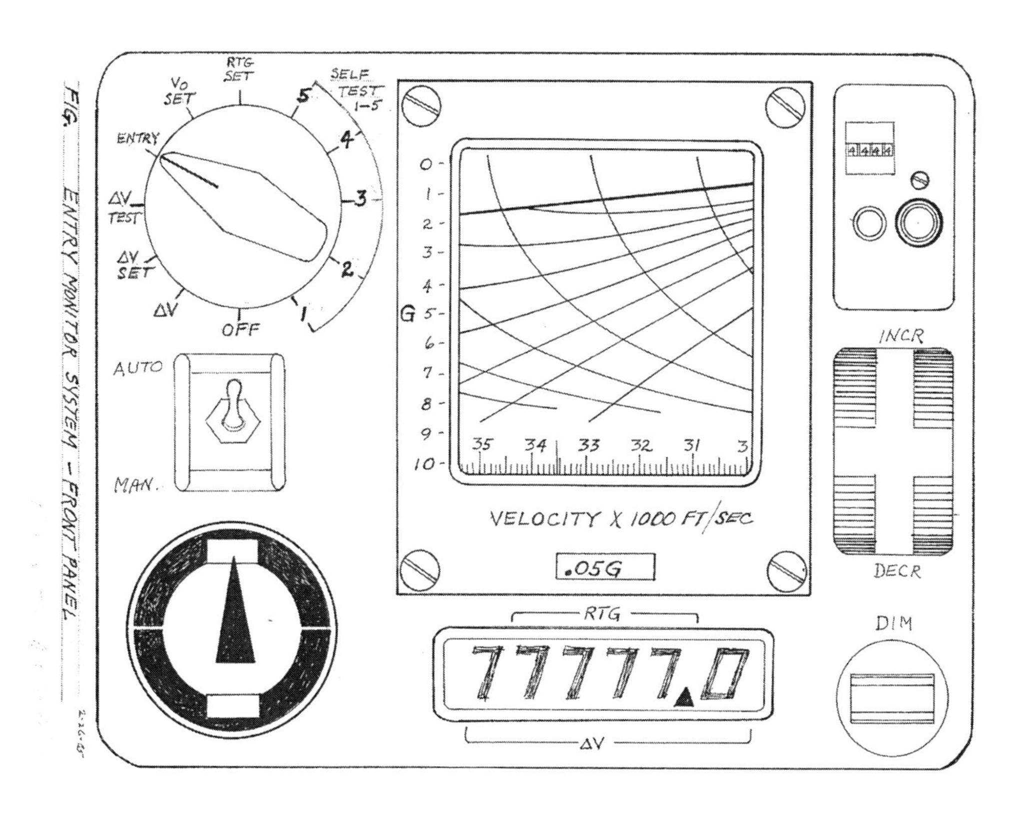

An artist drawing of the Entry Monitor System front panel used in a North American Aviation briefing (Courtesy of the Boeing Company).

The flaw was in the Entry Monitor System, part of the Block II Stabilization and Control System. The flight of the Apollo Command Module to the moon and back was done in the automatic mode by the onboard Command Module Computer (CMC) of the Primary Guidance, Navigation and Control System. As the reader will no doubt recall, the CMC was designed by the great Eagle Knight Dr. Ramon Alonso from MIT. In the early and mid-1960s, creating a digital computer to perform the onboard guidance, navigation and control functions for the Apollo moon mission was no easy task. Dr. Alonso designed the CMC using rope magnetic memory that he copied from the Australians, with 64 Kbytes of read-only memory (ROM) for the operating system and 8 Kbytes of random-access memory (RAM) for calculations.[1]

Throughout the whole flight to the moon, the three astronauts would be performing many different tasks, one of which was monitoring the performance of the Primary Guidance, Navigation and Control System to ensure that it was working properly. If the system should fail to perform its intended functions correctly, the astronauts would disengage the Primary Guidance, Navigation and Control System and replace it with its backup, the Block II Stabilization and Control System.

As mentioned before, the Block II Stabilization and Control System required the astronaut pilot to fly the Apollo capsule in the manual mode. The system provided the pilot with two Rotation Hand Controllers, one Translation Hand Controller and several flight indicator displays on the Command Module main control panel. One of these displays was the Entry Monitor System, which was used only during Earth reentry on a moon return trip. The Apollo capsule had to be captured by Earth's gravity and, in order to survive the heat of reentry, remain within a tubular corridor that got narrower as the capsule approached Earth. Drifting out of this corridor would mean certain death for the astronauts, either by being lost in space forever or by burning up during reentry. The Entry Monitor System provided the astronauts with a means of surviving a reentry in case the Primary Guidance, Navigation and Control System failed to hold the capsule inside the tubular corridor.

The Entry Monitor System had a window through which the Apollo crew could view the capsule's position within the corridor during the most critical time of the reentry period. The system displayed the gravitational force exerted on the spacecraft versus spacecraft velocity during reentry. The window display of the Entry Monitor System would immediately let the crew know whether the capsule had breached the upper or lower boundaries of the reentry corridor. If the Primary Guidance, Navigation and Control System did not immediately bring the capsule back inside the corridor, the astronauts could take over manual control of the capsule by engaging the Block II Stabilization and Control System, which was done by rotating the Translation Hand Controller T-handle clockwise. From that point on, by using the Rotation Hand Controller, the astronaut pilot would be able to manually fly the Apollo Command Module so that the Command Module stayed within the entry corridor, as displayed by the Entry Monitor System display.

The Entry Monitor System was manufactured by Autonetics and contained a scroll of Mylar film that went from the right side of the display to the left side by means of a take-up reel run by an electric motor on the left side. Mylar was a nonflammable plastic developed for the aerospace industry. In the manufacturing process, the back side of the Mylar scroll was coated with a liquid emulsion that was given time to dry before the scroll was installed in the Entry Monitor System. As the Mylar film scroll traveled across the display, a sharp metal tip located on the right side of the Entry Monitor System (held by spring pressure against the back side of the Mylar film) would scratch the dry emulsion from the film, thereby scribing the reentry path of the Apollo capsule, which continuously made the astronaut crew aware of whether they were inside the reentry corridor. The position of the sharp metal tip behind the Mylar film was driven by parameter data that showed the position of the Apollo capsule inside the tubular corridor.

The system flaw was identified by two events. First, on the Apollo 9 mission, the Entry Monitor System scroll failed to scribe for a short period during the time when the system was running prior to capsule reentry into Earth's atmosphere; later, the Apollo 10 Entry Monitor System scroll failed to scribe in a test of the scroll during initiation of the Entry Monitor System prior to Earth reentry. The doctors on the ground could tell exactly when the anomalies happened on Apollo 9 and 10 because the astronauts' heart rates showed a significant jump at the precise time that the scroll failed to record. When the Apollo 9 scroll failed to scribe for a short period, it was not automatically indicative of a problem, but the Apollo 10 incident raised a red flag. Something had to be done to ensure that nothing similar happened on Apollo 11 or any of the following flights.

My engineering notes from this time period indicate that work on this problem was carried out from May 29, 1969, to July 21, 1969. Although I was assigned to Carl Conrad's Stabilization and Control System unit and had my responsibilities there, I was

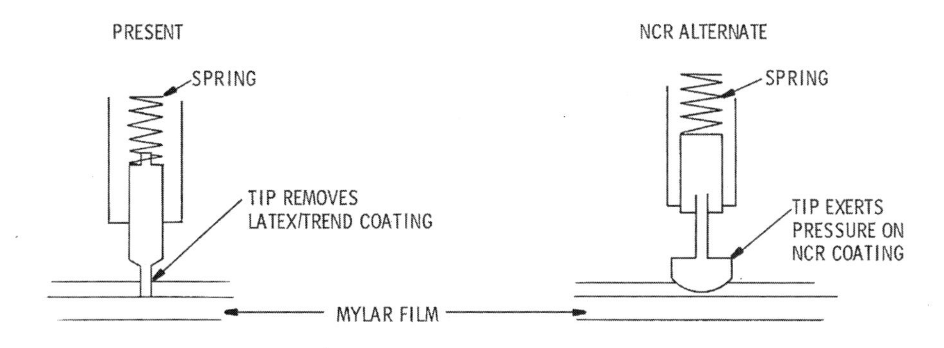

CHANGES REQUIRED FOR NCR ALTERNATE

- NCR SCROLL COATING
- STYLUS CHANGE - .260 SPHERICAL RADIUS TIP - NUB REMOVED
- NEW SPRING - 24 ± 5 OZ PRELOAD

This drawing shows the changes that had to be made to the Entry Monitor System to incorporate the NCR scroll (Courtesy of the Boeing Company).

called in to work on the Entry Monitor System problem despite never having worked on it before—no doubt because I had experience in designing and fixing systems and dealing with subcontractors. At the time that I was brought in to help solve the problem, the people involved from the Space Division were team leader Bill Fouts (Guidance and Control Department manager), Bill Paxton (Entry Monitor System unit supervisor), Thom Brown (Entry Monitor System lead engineer), Ken Saterlie (Entry Monitor System engineer) and Bob Epple (Guidance and Control Department assistant manager). Personnel from Autonetics were Jerry Shere (Entry Monitor System program manager), Norm Rassmussen (head of manufacturing), Bill Knox (mechanical project engineer), John Gressard (the Autonetics engineer who suggested that the National Cash Register [NCR] carbonless paper process could be adapted to the Entry Monitor System scroll and the most knowledgeable of the group about NCR carbonless paper), and Ben Reina (Entry Monitor System NASA system manager and another Eagle Knight).

After considerable investigation, we found that the problem stemmed from the consistency of the ingredients in the emulsion used to coat the back side of the Entry Monitor System Mylar scroll, which were not scientifically controlled and caused many scroll assemblies at Autonetics to be rejected because they did not meet the established acceptance tests. It was later discovered that the last step in the final acceptance of all Mylar scrolls was to call a person cited by name in the acceptance test procedure and have him scratch the emulsion of a coupon on the scroll with his fingernail to give the final aye or nay on whether the scroll would be accepted for Apollo flight. First, the scratching of a Mylar scroll coupon by a person for final acceptance was not quality control procedure that met Apollo NASA program requirement standards, so this approach had to be revised. More serious, however, was that one of the ingredients in the emulsion used to coat the back side of the scrolls was a dish soap sold by the Giant Food Market. The soap was being bought off the shelf at the grocery store—another example of disregarding established NASA requirements. Right around the time that the soap for the Apollo 9 and 10 scrolls was obtained, Giant Food Market had changed the ingredients used in the soap and of course degraded its quality, which in turn degraded the emulsion for the Entry Monitor System scroll. This process had to be replaced with an Apollo-accepted process, and I was brought into the team by Bill Fouts for the purpose of bringing this process to fruition with NCR using their carbonless paper process, which Autonetics hoped could be applied to create a better and more reliable Entry Monitor System Mylar scroll.

So, what was the National Cash Register, and why was this organization getting involved in the Apollo program? The following is what John Hanny, assistant manager of the NCR facility in Dayton, Ohio, told John Gressard, Thom Brown, and me on our first trip to NCR. The Dayton facility did not produce any paper; rather, it produced the solution that coated the carbonless paper. At the time, the facility had only about six employees, made millions of dollars a year and was very profitable.

The facility was opened when an NCR employee found a paper at the Stanford University library that showed possibilities for creating profitable products. This person's job was to visit different American universities to review the papers or theses written and published by master's and PhD candidates to see whether any of them contained discoveries or processes useful to NCR. The paper he found described how, when two solutions that do not blend (such as water and oil) are suspended in a mixture and the mixture is stirred for a long period of time, an emulsion is formed, composed of microscopic bubbles of one solution that microencapsulates the second solution. What you have at the end of the stirring period is an emulsion of millions of microcapsules of one of the solutions in the core of the microcapsules.

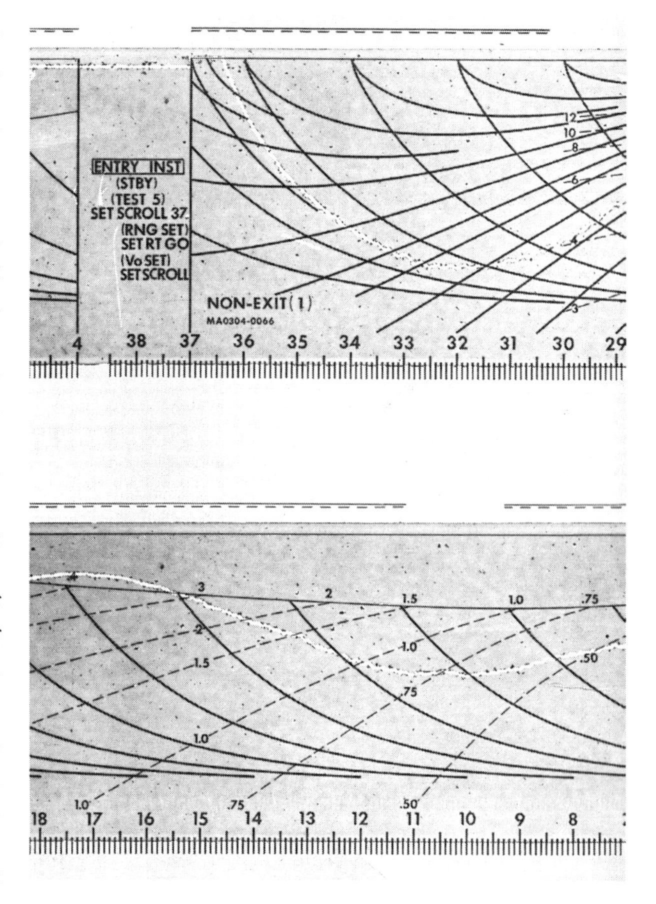

This is a copy of part of the Apollo 10 mission Entry Monitor System scroll that shows exactly what the astronauts were viewing during reentry (Courtesy of the Boeing Company).

NCR's process worked as follows: They filled two vats each with two un-soluble liquids. In the first vat, they poured a colorless un-soluble solution A that was suspended in a second solution. In the second vat, they poured a colorless solution B that was again suspended in the second liquid. The solutions in both vats were then stirred until one vat contained millions of microcapsules consisting of the A solution microencapsulated within the emulsion, while the second vat contained the same result with the B solution. The solutions A and B were the two ingredients that, when combined and mixed, chemically formed a dye of either blue or red. The clear A solution was the dye that determined the color, and the clear B solution was the co-reactant that chemically combined with the A solution to bring out the dye color.

In the NCR product intended for carbonless paper, the emulsion with the dye solution A and the emulsion with the co-reactant solution B remained separate, each in its own vat. The next step was to take a typical paper business form with a master top page and a second carbon copy page with a sheet of carbon paper between them. They discarded the carbon paper, leaving a master business form with a master top page and an attached copy page. The form was then opened out, with the master page folded back from the copy page that was now lying flat on a table. Then the back side of the master top page was coated with the emulsion with the microencapsulated co-reactant and allowed to dry. Next, the front side of the business form copy page was coated with the emulsion with the microencapsulated dye solution. When both pages were dry, the front page was folded back to its normal position. At this point, the back side of the master page coated with the co-reactant was in contact with the front of the copy page that was coated with the dye. With a ballpoint pen, they wrote on the master page, and with each stroke the pressure of the ballpoint pen broke the dye and co-reactant microcapsules that were in contact with each other. This action allowed the dye and co-reactant liquid to flow out and mix together to chemically form the red or blue color on the front of the copy page. Bingo, now we had NCR carbonless paper.

The plan that John Gressard and NCR envisioned was that NCR would combine the separate vats of microencapsulated co-reactant and dye solutions in a single vat without bursting the microcapsules. Then they would take the combined solution and use it to coat the back of an Entry Monitor System Mylar scroll. Instead of a sharp probe rubbing against the back side of the Mylar scroll as it moved from right to left (as was done in the old design), there would be a blunt probe making contact with the newly coated scroll that would have sufficient pressure to break the encapsulated microcapsules, causing them to chemically combine to form a red or blue line, thereby giving the astronauts a visual indication of the Apollo capsule's entry path so they could take action if the capsule breached the entry corridor and the Primary Guidance, Navigation and Control System did not immediately bring the spacecraft back inside the corridor.

As part of this project, I had to live with the problem at home in Downey, at Autonetics and at NCR in Dayton, Ohio, until there was a scroll produced that met all the Apollo requirements and could replace the old Entry Monitor System scroll. Although the first entry in my notebook was dated May 29, 1969, I know I started earlier than that because on that day our team held a meeting at Autonetics with Leo Krupp and Al Moyles (the division's two test pilots), Dave H. (the Autonetics test pilot), Bill Knox, Norm Rassmussen, John Gressard and me in attendance. Bill Fouts' team thoroughly reviewed a scroll that had been brought back from NCR. The Autonetics test pilot liked it but had the following comments:

1. The color of the scribed path on the scroll needs to be a deeper color.
2. The scribed path on the scroll needs to be wider.

(On my first trip to NCR, they had not yet successfully mixed the two independent solutions in one vat and thus could not possibly have coated a test scroll on May 29.)

I made numerous trips to NCR in Dayton to solve the problem of the failed Entry Monitor System scroll. Some trips were made for the purpose of setting requirements for the process of coating the scrolls with the NCR emulsion, establishing requirements for testing a coated scroll, and hand carrying Entry Monitor System scrolls to NCR to be coated with the emulsion. My last trip to NCR was to finalize the final design of the new coated scroll and sign off on all the specifications and test documents, accompanied by one of the company's buyers and Archie Hebert (pronounced "Ee-bear"), the reliability engineer assigned to the problem, to agree on the final purchase order to buy the coated scrolls from NCR.

Bill Knox, the Autonetics mechanical project engineer, conducted many tests on NCR coated scrolls at Autonetics to certify the scrolls for Apollo flight. I spent several days at Autonetics witnessing the tests that Bill was conducting. On June 2, 1969, there was a meeting in Bill Fouts' office with R.G.E. Epple, Bill Paxton, Thom Brown, and me to discuss a complete plan on how the team planned to get the new Entry Monitor System scroll with the NCR coating for Apollo 11 ready to present to George Jeffs (the Apollo program manager) on that same day. Bill Fouts' team plan included North American Rockwell's takeover of NCR management, which simply meant that I was supposed to be Mr. North American Aviation while I was at NCR and take over their management of the scroll activity, which was not a problem for me because I had done something similar at Autonetics on the AFRM 009 Control Programmer and the Mission Control Programmer for AFRMs 011, 017 and 020. Fouts' team plan also had to include the total program cost and an itemized list of what we received for this money. In addition, the team had to consider controlled tests on a backup scroll; Ken Saterlie had run uncontrolled tests, but controlled tests were needed. The Apollo capsule's velocity during abort entry was 38,500 feet per second, so the plan also had to make sure that it clarified whether there was a problem at that velocity. The team was required to list on the plan all the things that Autonetics found unsatisfactory with the scroll, along with the recommended changes. Bill Fouts announced that there would be a 2:15 p.m. conference call with Gerry Shere to discuss the plan of action for the combined effort between Autonetics and North American Rockwell Space Division. The team agreed to get together at 2:00 p.m. before the telephone call. The conference call with Autonetics took place in Bob Epple's office; attendees were Bill Fouts, Bob Epple, Bill Paxton, Thom Brown, and me. At Autonetics were Gerry Shere and Enus Fairchild. We all decided that we should hold a meeting at Autonetics the following day at 8:00 a.m.

On Thursday, June 3, 1969, Fouts' team attended the meeting at Autonetics, which lasted most of the day, attended by Bill Fouts, Bob Epple, Bill Paxton, Thom Brown, and me, as well as Gerry Shere, Bill Knox, and John Gressard from Autonetics. The composition of the two solutions that coat the Mylar scrolls was described in detail by John Gressard. Of the scrolls already coated, examples had been shown to Leo Krupp (NAR Space Division chief test pilot) and Al Moyles (NAR Space Division test pilot) on May 28, 1969. Autonetics had three scrolls, one shown to Krupp and Moyles, plus two to be taken to Houston for centrifuge testing by NASA engineers. There were three types of scrolls of three different colors, with each color being defined by its technical name. As agreed, five scrolls would be sent to NASA at Houston, and the Space Division would also ensure that the scrolls met the Apollo flammability requirements, the smell (odor) requirements, and out-gassing requirements. NCR used a special technique to apply the coating to the scrolls and could do 30 scrolls in two weeks. Thirty NCR scrolls could be provided to Autonetics for $10,000.

The next topic we discussed was the accountability and location of all the Entry Monitor Control assembly units and the Entry Monitor Scroll assembly units. Autonetics reported how many units of the two types were available and where they were presently located. Bill Fouts' team needed to know this information to ensure that there were enough units to support the Apollo program. The total cost estimated by Autonetics was given to us as approximately $30,000, which included the 30 NCR scrolls and the Autonetics engineering and test effort of the scrolls, with some overtime included and possibly four trips to Dayton, Ohio. The meeting was adjourned, and Fouts' team returned to Downey.

From June 9 through June 27, 1969, I met with Project Engineer Bill Knox at least five times for the purpose of discussing the environmental tests and scheduling for them to qualify the NCR scrolls for the Apollo program. The following paragraphs will discuss some details of these meetings only because they contribute something to the solution of the problem and provide some behind-the-scenes activity. During the meeting between Bill Knox and me on June 9, besides the scroll environmental tests, we talked about the controls to be negotiated with NCR during my upcoming trip to Dayton. At the end of that meeting, Bill and I agreed on nine program controls to be negotiated with NCR for the qualification tests of the Apollo scrolls.

My meeting at NCR in Dayton was a two-day meeting on Thursday, June 12, and Friday, June 13, followed by another meeting on Monday, June 16, 1969, at Autonetics to review all that transpired in Dayton the previous week. Attendees at the meeting in Dayton on the first day were Joe Sakon, the NCR facility manager; John Hanny, the NCR assistant manager; Nick Mannali, one of NCR's research engineers; Bob Haines, another NCR research engineer; John Gressard from Autonetics; and myself from NAR Space Division. NCR personnel described the process they used to coat the Mylar scroll film and the ingredients used in the solutions. The significant point was that once the solutions were mixed, the Mylar film had to be coated within 2 minutes; otherwise, the coating came out pink. John Gressard and I were given a list of the samples of ingredients that NCR was going to provide to Autonetics and the pieces of coated film, along with any results of different tests performed on the completed product. John and I elected to review all the material given to us when we returned to California and see whether it met Apollo program requirements. At the end of the meeting, NCR presented John and me with a schedule of scrolls to be used for testing by Autonetics and the different types to be delivered, which John and I agreed was in line with Autonetics' required dates. John Hanny was informed that Dick Brooks, the buyer, would write a letter to NCR to get the test scrolls delivered out of NCR on June 16, 18, 20, and 24 prior to receipt of a purchase order. And, of course, NCR wanted to know when the Apollo program needed the flight scrolls, which John and I could not answer at the time, but we told them the date would be provided later. Other questions that were asked, which could not be answered, were "Will the entry monitor scroll assembly acceptance test start on July 1, or will it start earlier? If earlier, what date?"

On the second day of the meeting, John Gressard and I presented a briefing to NCR management and to the personnel working on the NCR Entry Monitor System scroll. The presentation was well received by all, and a discussion and questions period followed. At 1:30 p.m., the letter from Dick Brooks was signed and delivered to NCR in lieu of a purchase order. Bob Haines agreed to coat some scrolls for John Gressard to hand carry to Autonetics. On Monday, June 16, John and I held our meeting at Autonetics to review the trip to NCR at Dayton. Other attendees were H.P. Buist and A.T. Tarbell, Jr. The scrolls that were hand carried by John to Autonetics on Sunday, June 15, were made available for testing.

Besides going to NCR in Dayton for meetings and attending meetings at the home plant in Downey and Autonetics, I also visited Autonetics to witness the environmental testing that Bill Knox and his crew were conducting on the NCR scrolls to certify them for Apollo space flight. On Tuesday, June 17, I went to Autonetics with Ben Reina, the NASA Entry Monitor System manager, who was in from the Manned Space Flight Center in Houston to witness the scroll vibration tests, accompanied by John Gressard and Bill Knox. The two axes vibration tests of the scroll we four witnessed were both good, with an acceptable scroll trace being scribed on the film during vibration test, and so Ben, Bill and I signed the test procedure to make the test official.

Ben and I returned to Autonetics in Anaheim the following day, and early that morning Frank Van Derwalker was given a 26-inch section of one of the scrolls for a vacuum test we called the "Ben Reina Test" because Ben had devised the procedure and requested that it be run. Ben and I were there that day, monitoring all the activity of the scroll temperature tests until approximately 9:30 p.m., when the team ran a successful test to ensure that the unit would scribe the entry path at 150 degrees. The next day, at approximately 9:00 a.m., the temperature chamber was opened, and a post-temperature test was completed under ambient temperature conditions on the test scroll, and the results looked good. Ben Reina returned to Houston the following day.

I returned to Autonetics on Wednesday, June 25, 1969, to witness the humidity tests. The tests ran successfully; then the team ran a trace on the NCR Mylar scroll as a post-humidity test, which went well. At 6:30 p.m. on Thursday, June 26, Bill Knox's test team began to run the post-humidity leak tests on the Qualification Test Entry Monitor Scroll Assembly. The team found that the scroll would not drive forward or in the reverse direction. This was a new problem that the Apollo program could not afford because the launch of Apollo 11 was less than a month away. The cover was removed from the Entry Monitor Scroll Assembly at 7:35 p.m.; several problems were found that the test team tried to correct, but they had no success. It was decided that the next step would be to check the mechanical scroll drive assembly. At 8:45 p.m., a test of the scroll drive assembly showed low parameter specification readings. The test team devised a plan that I approved for the solution of this problem. The Entry Monitor Scroll Assembly NCR scroll-fail-to-reverse problem was a critical juncture in the attempt to get a reliable scroll for Apollo 11, and so its solution became a high priority.

On Friday, June 27, 1969, less than a month before the moon flight of Apollo 11, there was a meeting held at Autonetics regarding the NCR scroll reversal problem. The attendees were Gerry Shere, Bill Knox, Rick Harris, Kass Bulota, John Gressard, Norm Rassmussen, Ken Saterlie, Bob Epple, Bill Fouts, Thom Brown, and me. Bill Knox and I briefed everyone else on the various tests that had been run the day before and all the test data retrieved. Autonetics agreed that they would quickly come up with a solution to the scroll-fail-to-reverse problem.

Bob Epple, Bill Paxton, Thom Brown, and I met in Bob Epple's office on Wednesday, July 2. The team discussed a mission rule for Apollo 11 that required a demonstration of the operation of the new NCR scroll with a backup scroll during the Kennedy Space Center ground test. A demonstration of the NCR scroll given to the astronauts resulted in a comment from astronaut Joe Engle that the line width was too narrow. Our team was sure that the stylus Autonetics was using produced a wide line, and we surmised that the system used for the demonstration probably did not have the correct stylus pressure. Other topics under discussion were the requirements that the team wanted to impose on Autonetics regarding the data that they should collect for Space Division when testing the NCR scroll for installation into an Entry Monitor System. The team listed three parameters that had to be recorded during the tests and had Autonetics provide the data that the Space Division would request from Kass Bulota. The team also had a fourth requirement to request—namely, for Autonetics to run a final test pattern test on each Entry Monitor Control Assembly. At the close of the meeting, Bob Epple asked me to call John Hanny at NCR in Dayton and get a status update on the scroll program.

My last meeting with Bill Fouts was on Monday, July 21, 1969, at which time we discussed the upcoming Entry Monitor Scroll Assembly dry run test at 9:10 a.m. on Wednesday with the NCR scroll. On Friday, July 25, Autonetics would brief Fouts' team on the fail-to-reverse problem fix. Bill also mentioned an Entry Monitor System briefing he was to give for Spacecraft 108 (Apollo 12) Flight Readiness Review at the NASA Manned Spacecraft Center in Houston.

The last trip I made to NCR in Dayton was for a couple of days; I was accompanied by Archie Hebert, the reliability engineer assigned to the Entry Monitor System problem. Archie, a retired engineer from the Los Angeles Fire Department, was one of very few Black engineers whom I had encountered during my career, as well as a good friend and colleague. The purpose of our trip to NCR was to finalize the engineering and quality control requirements for the Entry Monitor System scroll for Apollo 12 and subsequent spacecraft. Because I was now working on the application of the new Mylar scroll for Apollo 12, I had no knowledge of what solution for the Apollo 11 Entry Monitor System scroll had been agreed to by NASA management and Bill Fouts in conjunction with other Apollo program higher managers, so I cannot comment on that.

Archie and I met with Joe Sakon, John Hanny, Nick Mannali, and Bob Haines for two days while they went through all the engineering details of how the scrolls were processed in the facility and the stringent quality controls in place to ensure that the Apollo program was getting reliable scrolls. Archie and I were also presented with all the process specifications used, along with the different tests imposed on all scrolls and the data that would be provided to NAR Space Division by NCR. We reviewed all the process specifications and the test specifications to make sure every requirement had been met, and where we found paragraph wording that we wanted changed, we were quickly accommodated. At the end of the two days, Archie and I had agreement on every engineering and reliability document, so NAR Space Division sent a buyer to sign the finalized contract. We signed all the engineering and quality control documents, along with all other agreements for NAR Space Division engineering.

Archie and I had a late afternoon flight that day, so before we drove to the airport, I placed a telephone call to Bill Fouts and asked him to get Bob Epple, Bill Paxton, Thom Brown and whoever else from engineering he wanted so I could tell them what Archie and I had signed as an engineering agreement with NCR. As soon as Bill had everyone he wanted in his office, I proceeded to relate everything that Archie and I had reviewed and signed, with details of the scroll build process and test details and the quality control requirements placed on the scroll acceptance process. One of the questions asked was whether there was a scientific quantitative acceptance test on each scroll to ensure that there was consistency in the scroll quality. I answered that there was an acceptance opacity test of the dry emulsion of each scroll with a nominal value and plus and minus limits, which satisfied everyone in Fouts' office. There were many more questions asked, which I answered to everyone's satisfaction. At the end of the telephone call, I told Bill and everyone in his office that what I had described were contractual agreements with NCR and that Archie and I were heading to the airport within one hour, so if anyone had any disagreements, they had better call back at the phone number I had given them before we left; after that, it would be too late and any change would require a contract renegotiation.

Archie and I waited next to the phone, looking at our watches and expecting a telephone call from Downey at any minute to tell us not to leave until some suggested changes were made to the contract. At the expiration of the hour, no one from Bill Fouts' department at the NAR Space Division in Downey called us back, so Archie and I headed for the airport to catch our flight.

23

We Land on the Moon and a Few More Eagle Knights

Besides the Entry Monitor System problem, there were other issues to resolve before Apollo 11 could be launched; once that goal had been achieved, the six Apollo Command Modules that followed would be free and clear for a moon landing mission. According to my notes, there were at least six more problems that were solved prior to the Apollo 11 launch.

According to the unit organization chart that I have (dated about two weeks after Archie Hebert and I returned from Dayton, Ohio), Carl Conrad's Guidance and Control System Unit consisted of thirteen engineers and one secretary.[1] The secretary was Jerri Sutton, who had been with the company several years. Carl had organized his unit into six subunits, some of which included only one person. The biggest subunit was Guidance and Control Evaluation and Flight Support, with three principal guidance and control engineers whose names were Al H. Sohler (responsible for Apollo 8), Jim E. Roberts (responsible for Apollo 9), and Bob O. Zermuehlen (responsible for Apollo 7 and Apollo 10). The second subunit was Entry Monitor System Design, with Thom L. Brown as lead engineer and Harry S. Markarian, Ken R. Saterlie, and Fred O. Wadman under his leadership. The third subunit was Stabilization and Control System Electronics Design, with George Cortes and Mert T. Stiles as his lead engineer. Then there was Stabilization and Control System Displays and Controls, with Chuck W. Markley and me as lead engineer, followed by the Design Qualification Certification and Configuration subunit with Bernice W. Johnston. The last subunit was Manager's Office Support, which consisted of Milt W. Swan.

Around late July or early August, Carl Conrad told me that he had inherited the Entry Monitor System along with Bill Paxton, Thom Brown, and the remainder of the Entry Monitor System personnel. Sometime after Bill Paxton came to work for Carl, Bill told Carl that if there was a layoff at some later date, he would not mind getting let go. Shortly afterward, Carl laid Bill off, something that Carl had been in favor of for a long time. Bill told Carl that he had meant later, but it was too late; my old supervisor Bill Paxton was gone, and I shed no tears because he was without doubt the worst supervisor I ever had.

So, now to get back to the launch of Apollo 11. The Apollo program was finally going to get a chance to land two American astronauts on the moon. The success of Apollo 10 had been the final key to giving Apollo 11 the go-ahead to make a moon landing. The Apollo 11 Command and Service Modules were delivered to NASA by NAR Space Division on Wednesday, January 22, 1969. At delivery, this spacecraft was destined for possible use in a lunar landing mission.

Al Kehlet, assistant program manager for the Command and Service Modules of the Apollo 11 spacecraft for almost five years, was assigned to lead the team responsible for getting the Apollo 11 spacecraft through NAR Space Division in preparation for delivery to NASA.[2] He had an impressive record that dated back to the Mercury and Gemini programs. Al Alcantar (another Hispanic Eagle Knight) was the project engineer joining Al Kehlet on the Apollo 11 spacecraft for the NASA Manned Spacecraft Center's resident office in Downey. They were the key people who made the delivery of the spacecraft for Apollo 11 to NASA possible.

A photograph on page 3 of the February 7, 1969, *Skywriter* showed the Apollo 11 Command Module and Service Module being uncrated after arrival at Kennedy Space Center in the week of February 7, 1969. The uncrating was followed by the connection of the spacecraft with the sections of the Saturn S-5B booster.[3] More information on the rocket to the moon launch vehicle was reported in the June 20, 1969, *Skywriter* (now known as the *North American Rockwell News*), which announced that the Flight Readiness Test was underway for Apollo 11. On July 11, 1969, Columbia and Eagle (the code names given to the Command and Lunar Modules, respectively, by the astronauts) were prepared for launch. Apollo 11 was in the final countdown for lift-off at 6:32 a.m. on July 16.[4] The commander of the spaceship (or, as a Russian newspaper would call him, "the czar of the spaceship") was Neil Armstrong; Buzz Aldrin was the Lunar Module pilot, and Michael Collins served as the Command Module pilot. Since Michael Collins was the Command Module pilot, he was to stay on board the Command Module as it orbited the moon while Neil and Buzz descended to the surface in the Lunar Module to make history as the first two humans to set foot on the moon.

Apollo 11 was launched July 16, 1969, and on July 20 made a miraculous landing on the moon. It returned to Earth on July 24. As stressed throughout this book, nothing about this mission could be considered simple. Sending a manned spaceship on a voyage to the moon and back is indeed a risky and extremely difficult and technically complicated task. To the skeptics who do not believe that we landed on the moon, it can only be said that those employees of companies who worked on the Apollo program know that on July 20, 1969, Neil Armstrong was the first undisputed human to put his footprint on the moon.[5]

So, who were these three brave men who sat atop a rocket loaded with explosive fuel, going to a place they had never been before and risking the possibility of not coming back? The astronauts inside the Apollo 11 Command Module were three very experienced pilots and exceedingly qualified to fly the two complex machines, Columbia and Eagle. They had logged an aggregate of 762 hours in space and 10,316 hours in aircraft (8,273 of them in jets).

Commander Neil Armstrong was a former test pilot for NASA and North American Aviation; he flew the X-15 to an altitude of 400,000 feet and a speed of 4,000 miles per hour. He was 38 years old at the time of the moon mission, having been born in Wapakoneta, Ohio; he had earned a Bachelor of Science degree from Purdue University and later took graduate work at USC. Armstrong was a civilian astronaut, having served as a naval aviator from 1949 to 1952, and flew 78 combat missions during the Korean action. As command pilot for the Gemini 8 mission, he performed the first successful docking of two vehicles in space.

Command Module pilot Michael Collins was 38 years old and an Air Force lieutenant colonel who had earned a Bachelor of Science degree from the U.S. Military

Academy at West Point. He served as an experimental test officer at the Air Force Test Center, Edwards Air Force Base, in California. During the Gemini program, he flew with John Young in Gemini 10 and completed 89 minutes of extravehicular activity, which made him a spacewalker.

Lunar Module pilot Edwin (Buzz) Aldrin was an Air Force colonel, 39 years old, who had earned a Bachelor of Science degree from West Point and a Doctor of Science degree from MIT. He flew 66 combat missions in Korea in a North American Aviation–built F-86 jet fighter aircraft and was credited with two aircraft kills of Russian-built MIG-15 jet aircraft and destroying a third. Later, in Germany, Aldrin piloted the North American Aviation–built F-100 jet aircraft. He flew in the Gemini 12 space capsule with James Lovell, which brought that program to a successful end. During the Gemini program, he set a record for extravehicular activity of five and a half hours.[6]

Apollo 11 accomplished what no human had ever attempted before, but the Apollo program accomplishments prior to this mission were in themselves impressive and included four Apollo Command Modules manufactured by the Space Division of North American Rockwell in Downey, California, containing 7,999,930 perfectly working parts that traveled more than 9 million miles through space. Each Command Module had nearly two million functional parts, excluding wiring and structural parts. Apollo 7, 8, 9 and 10 accrued two trips to the moon and back, 41 lunar orbits, and 318 Earth orbits (roughly 9,470,000 miles) in a total of 840 hours (equivalent to 35 days). On the manned flights prior to Apollo 11, approximately 70 parts had failed, none of which had jeopardized the safety of the crew or the success of the mission. All this was accomplished in less than one decade and with 1960s technology and a slide rule.

Modern readers cannot imagine the excitement that everyone felt and the hype that permeated the atmosphere at the Apollo facility in Downey prior to the Apollo 11 flight. Every day everyone was reminded of it in conversations and in the various published newspapers. Here is a letter written by our division president Bill Bergen that appeared in the company newspaper five days prior to the Apollo 11 launch:

> The Apollo 11 mission, like most other great events in history is essentially a story of people. When the late president Kennedy proposed the lunar landing program, he observed that "... in a very real sense, it will not be one man going to the moon ... it will be an entire nation. For all of us must work to put him there."
>
> Through the eight years leading up to the launch next Wednesday, approximately 300,000 persons in 20,000 companies throughout the United States have been striving with NASA to ensure that the first footprints on the moon would be made by an American.
>
> Because of our efforts of building the Saturn S-II and the Apollo Spacecraft Command and Service Modules, no company has shouldered a greater responsibility than the Space Division. To no other group of industrial employees can it be more truly said: This is your mission. A part of the product of your labor will be launched with the three valiant astronauts. I am proud to be a part of the team and I am proud of you. We'll all be hoping, many of us will be praying, for the success of the mission. And through our continued efforts, we'll want to make sure that this lunar landing is only the beginning of space exploration for our country.
>
> /s/ William B. Bergen
> President

The July 11, 1969, issue of the North American newspaper, the *Rockwell News*, included a full-page letter, which appears below in its entirety.

America
is about to put men
on the moon.

Please read this before they go.

Perhaps the best way for anyone to try to understand the size of such an undertaking is not for us to list the thousands of problems that had to be overcome, but for you to go out in your back yard some night, and try to imagine how you'd begin, if it were up to you.

But our reason here is not to talk about the technicalities of the Apollo project. Rather, it is simply to ask you to think, for at least one brief moment, about the men and women who have applied their heads and their hearts and their hands—and a good many years of their lives—to putting a man on the moon.

Many of these people have worked for less money than they could have made in other places, and it is safe to say they have worked through more nights and weekends and lunch and dinner hours than they would have anywhere else.

And the astronauts, the brave men who will fly again down that long, dark, and dauntless corridor of space, this time to set foot—to walk upon the surface of the moon—they know the price that's often paid in setting out for lands uncharted. They know the price their fathers' grandfathers paid just to walk across the wilderness of America for the first fifty years.

For a long time now, we have been involved with the people who are the thinkers and the designers and the builders and the pilots of America's man-to-the-moon dream, of America's man-to-the-moon determination. We have worked with them, eaten with them, lived with them.

Yet our appreciation and admiration for them continues to grow each day—for their energy, for their imagination, their confidence, for their patience, for their resourcefulness, for their courage.

We ask you, in the days ahead as we wait for the big one to begin, to understand this fantastic feat for what it is and to put it in proper perspective, a triumph of man, of individuals, of truly great human beings. For our touch down on the moon will not be the product of magic, but the gift of men.

In James A. Michener's novel, "The Bridges at Toko-Ri," an American Admiral stands on the deck of his carrier early one morning and ponders the subject of his brave men. And thinking to himself, he asks a question of the wind which we believe all of us should ask of the men who will finally make it to the moon and of the men who got them there: "Why is America lucky to have such men? ... Where did we get such men?"

The ending of this letter is true to the point and includes the Eagle Knights. I was in awe of the people I worked with and the astronauts I met during the Apollo program. There were thousands of employees at the Space Division in Downey, California, and what stood out in all the people whom I met and worked with was that even though we were all of different ethnic cultures and different political and religious beliefs, we were all descendants of immigrants and shared one goal: to put a man (an American) on the moon by the end of the decade; such an achievement could happen only in America.

The Apollo mission was one of the most important events of the 20th century, and people knew it, so consequently Florida had a crowd problem. At least 1,000,000 visitors were expected to be in Florida to witness the Apollo 11 launch. All motels and hotels in Brevard County—home of the Kennedy Space Center—had been reserved since before the Apollo 10 flight. Because there was no more room in the inns, most companies started booking their VIPs in Vero Beach (50 miles south of the spaceport) and Daytona (50 miles north). The NASA Public Affairs Office even requested that private citizens with spare rooms contact them for possible rental agreements.[7] The local sheriff expected traffic to be a problem, causing chaos on the roads that were not meant to support the large volume of cars expected. Many local towns bordering on the Banana and Indian Rivers, across from the launch pads, cleared areas along highways for parking, with some towns even setting up speakers along the highway to transmit NASA commentaries to the crowds.[8]

In preparation for the mission, Norm Casson, manager of Spacecraft Checkout, Command Service Module Test Operations (and one of a few Black managers at the Space Division), dispatched two of his teams to Hawaii so they could make the Command Module safe for post-flight analysis after splashdown. This task consisted of propellant system draining and cleaning, deactivating pyrotechnic devices, and sealing radiation sources. In all, there were some 5,000 steps, and the "safing" was not a job for the inexperienced. The highly motivated men of Casson's teams had been performing these operations on recovered Command Modules since the unmanned test flights in 1966 starting with AFRM 009 (which, if you recall, was the first flyable Apollo capsule to go into space that was flown by the Control Programmer). One unit was composed of team leader C.E. McKim, H.F. Shimizu, C.H. Burch, F.A. Schmidt, and Eagle Knight Lino Salazar. The second team consisted of five other members, and there was also a backup team of six members.[9]

On July 16, 1969, Apollo 11 blasted off the launch pad and, after one and a half Earth orbits, reignited the S-4B booster engine to break away from Earth's gravity and head for the moon. Roughly a half hour later, the Command Module–Service Module combination separated from the S-4B booster and docked with the Lunar Module that was still attached to the booster. After the docking was completed and the Command Module (with the attached Lunar Module and Service Module) pulled away from the booster, the S-4B booster was ejected into solar orbit while the astronauts and their spacecraft continued toward the moon. As the Apollo spacecraft moved farther away from Earth, the pull of Earth's gravity diminished; likewise, the moon's gravitational pull increased the closer the spacecraft approached, causing it to pick up speed. The Service Module propulsion engine was reignited when the spacecraft was behind the moon to slow down the spacecraft and place it into lunar elliptical orbit. Later, the Service Module propulsion engine was reignited a second time to place the Apollo spacecraft in a circular orbit approximately 60 miles above the moon's surface.

From the circular orbit on Sunday, July 20, after Neil Armstrong and Buzz Aldrin entered the Lunar Module and separated from the Apollo spacecraft, they ignited the Lunar Module descent engine and headed for the surface of the moon. They landed safely on the Sea of Tranquility with only 30 seconds of fuel left in the descent stage of the landing craft. An estimated half billion persons heard Armstrong announce with cool objectivity at 1:15 p.m., "Houston.... Tranquility Base here.... The Eagle has landed." These were the first words uttered from the moon. The first one to leave the Lunar Module and step on the moon was Neil Armstrong, uttering his famous words about one small step and one giant leap that are now part of history. Buzz Aldrin followed to be the second man to stand on the moon.

The two astronauts remained on the lunar surface for a little over 21 hours, performing experiments and collecting moon rocks and material to bring back to Earth. Before they left the surface of the moon, they left three mementos behind. First, Commander Neil Armstrong unveiled a plaque attached to the descent stage of the

Lunar Module. The plaque was signed by President Richard Nixon and the three Apollo 11 astronauts. It bore images of the hemispheres of the Earth and the following inscription:

HERE MEN FROM THE PLANET EARTH
FIRST SET FOOT UPON THE MOON
JULY 1969, A. D.
WE CAME IN PEACE FOR ALL MANKIND

The second memento was a one-and-a-half-inch disc bearing messages of goodwill from heads of state of many nations. The third was the flag of the United States of America, erected by Neil Armstrong while Buzz Aldrin took photographs. Planting the flag was meant to serve as a symbol of the first time humanity had landed on another celestial body, rather than a territorial claim by the United States.

During the final telecast from space late Wednesday, Armstrong said, "We'd like to give special thanks to all those Americans who built the spacecraft, who constructed, designed and tested them and put their hearts and all of their abilities into the craft. To these people tonight we give a special thank you and to all the other people listening and watching, God bless you and good night from Apollo 11."

On July 24, the NA Rockwell–built Command Module Columbia, with the three astronauts on board, splashed down at 9:50 a.m. Pacific Daylight Time in the Pacific

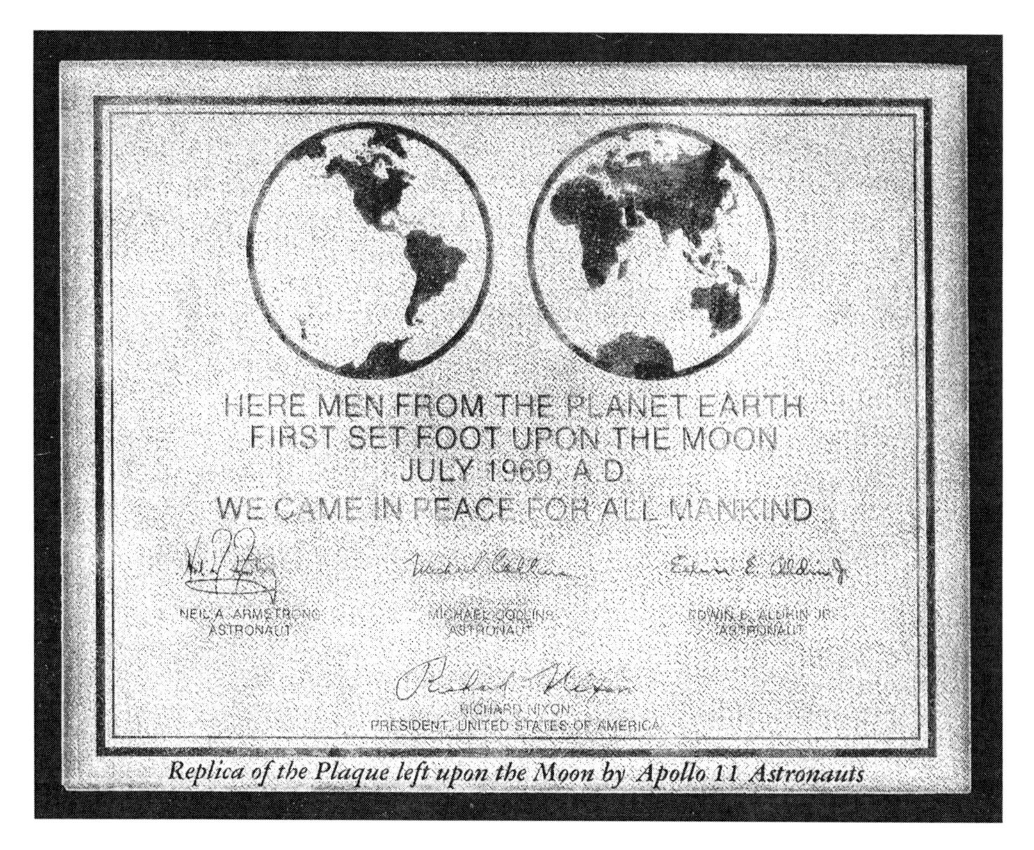

A replica of the plaque that was left on the moon by the Apollo 11 astronauts (author's collection).

Ocean, 912 miles southwest of Hawaii. They were picked up by the USS *Hornet* while President Nixon, on board the vessel, witnessed the operation. The three astronauts were quarantined in a vehicle called the Mobile Quarantine Facility (normally referred to as the MQF) on board the ship. The Apollo 11 crew brought back around 70 Earth pounds of moon rocks and soil material that were distributed to the appropriate facilities for future investigation.

Employees in Downey, California, eagerly awaited the return of the Apollo 11 Command Module, and on August 13 it arrived. Naturally, the Apollo 11 Command Module was on display for employees and their families. I took my family to view the Apollo 11 capsule, and even took pictures, because I wanted my family to be aware of what had been accomplished and that I had contributed to this historical event by helping to design some of the systems that were installed in the Apollo 11 capsule, giving up five years of my life to this project. Everyone at the Space Division in Downey was keenly aware of what had been accomplished, and the two top executives of North American Rockwell, W.F. Rockwell, Jr., and J.L. Atwood, expressed it clearly in a letter that appeared in the *North American Rockwell News* on July 25, 1969:

> This is a proud moment for mankind and for North American Rockwell and its employees. We have landed man on the moon and successfully returned him to Earth.
>
> We share with you an overwhelming sense of pride in being a major part of a magnificent industrial team which produced the systems for this historic expedition. These systems, for decades to come, will symbolize our nation's ability and willingness to accept challenges and solve major problems.
>
> We are confident that they also symbolize the spirit within North American Rockwell that will bring us many future successes.
>
> Congratulations for your contributions to this historic event.
>
> W.F. Rockwell, Jr.
> J.L. Atwood
> Chairman of the BoardPresident
> and Chief Executive Officer

Photograph taken by the author of the Apollo 11 capsule after returning from its mission to the moon (author's collection).

All employees at the Downey facility celebrated our achievement of finally putting men on the moon, and for a while we were relatively relaxed, but it did not last long because we still had six more moon landing missions to complete. We eventually got off cloud nine and went back to work.

Apollo 11 after returning from its mission to the moon (author's collection).

Apollo 11 after returning from its mission to the moon (author's collection).

The National Aeronautics and Space Administration
presents the
Apollo Achievement Award
to
J. DE LA ROSA

In appreciation of dedicated service to the nation as a member of the team which has advanced the nation's capabilities in aeronautics and space and demonstrated them in many outstanding accomplishments culminating in Apollo 11's successful achievement of man's first landing on the moon, July 20, 1969.

Signed at Washington, D.C.

ADMINISTRATOR, NASA

The Apollo certificate awarded to the author by NASA (author's collection).

I worked on the details for the Delta Qualification Test for the Translation Hand Controller Abort Redundancy redesign with Steve Scarborough of Honeywell. In addition, there were problems with some displays and controls and the gyro assembly that needed my attention.

July passed into August, and now Orville Littleton (the NASA Block II Stabilization and Control System manager), Honeywell and I were debating whether to use a refurbished Translation Hand Controller or a non-refurbished controller for the Delta Qualification Test. Orville favored the latter option. Project Engineer Chuck Moosbrugger of Honeywell and I kept working on the integrated chip problems.

The following names were published in the August 1, 1969, *North American Rockwell News*: "Apolonio Ramirez, E.R. Valles, J.R. Jaramillo, L.B. Olvera, William Cendejas, W.T. Sanchez, and R.R. Hernandez are members of the crew that attained a perfect Saturn S-II circumferential weld for the second time."

During the last week in July 1969, after the moon landing and the return of Apollo 11, I was asked by Carl Conrad to report to the Seal Beach facility for a new temporary assignment, but I reminded Carl that I was already scheduled to go on vacation from August 8 to September 2 and could not possibly fulfill the assignment. Carl sent Mert Stiles instead. I started my vacation with some apprehension because even though there were still six more moon missions pending, I thought that maybe when I returned to work, I was going to be reassigned to one of the several new inhouse research programs, even if only on a temporary basis.

On my return from vacation, I discovered that the special project Mert Stiles was assigned to in my place was the space station. Regardless, I started to ease back into my work with the Honeywell folks in Minneapolis.

The following names were published in the August 29, 1969, *North American Rockwell News*: "Anita Rosales of PRIDE Administration is accepting applications for the contest to name the Apollo 12 command module. Jorge Diaz is supervisor of Laboratories and Test's Chemical and Thermal Analysis group of laboratories where new space materials are tested to extreme hot and low temperatures to aid Space Division Engineers to consider the behavior of materials when designing spacecraft. Ray Sena of Procurement Quality Assurance won the men's tennis single title for the third straight year in the 1969 NR Space Division Tennis Tournament." In addition, the September 5 issue of the *Rockwell News* mentioned these individuals: "Fred Rodriguez, Robert Arebalo of Autonetics and Manny Alvarez, Hank Martinez, Phil Padilla, Jake Alarid, Joe Gomez, and Ted Garcia of Space Division are members of YITM and are shown in a photograph on page 4 of the company newspaper planning activities to combat the 'dropout' problem in the local schools." And then, on October 17, 1969: "H.L. Alcantar a member of the NASA Apollo Resident Project Office is one of 16 Apollo program members to receive certificates of appreciation from NASA's George Low for their contribution to the Apollo program."

The author at the Astronauts Memorial at Kennedy Space Center in 1995 (author's collection).

For the first few days after my return from vacation, I worked on solving some problems, one being the Apollo 14 Rotation Hand Controller, which had a damaged cable that had been pinched by a foot pan. I went to Building 290 and went through the process required to enter the Apollo 14 capsule to inspect the extent of the damage. There was a slight tear on the Fluorel cover, which did not really hurt the cable. The zipper tubing was on the cable and was damaged in the same spot. No action to repair was required.

The last two entries in my notes concerning my association with the Apollo program were made on September 8 and 11, 1969, and involved telephone calls with Project

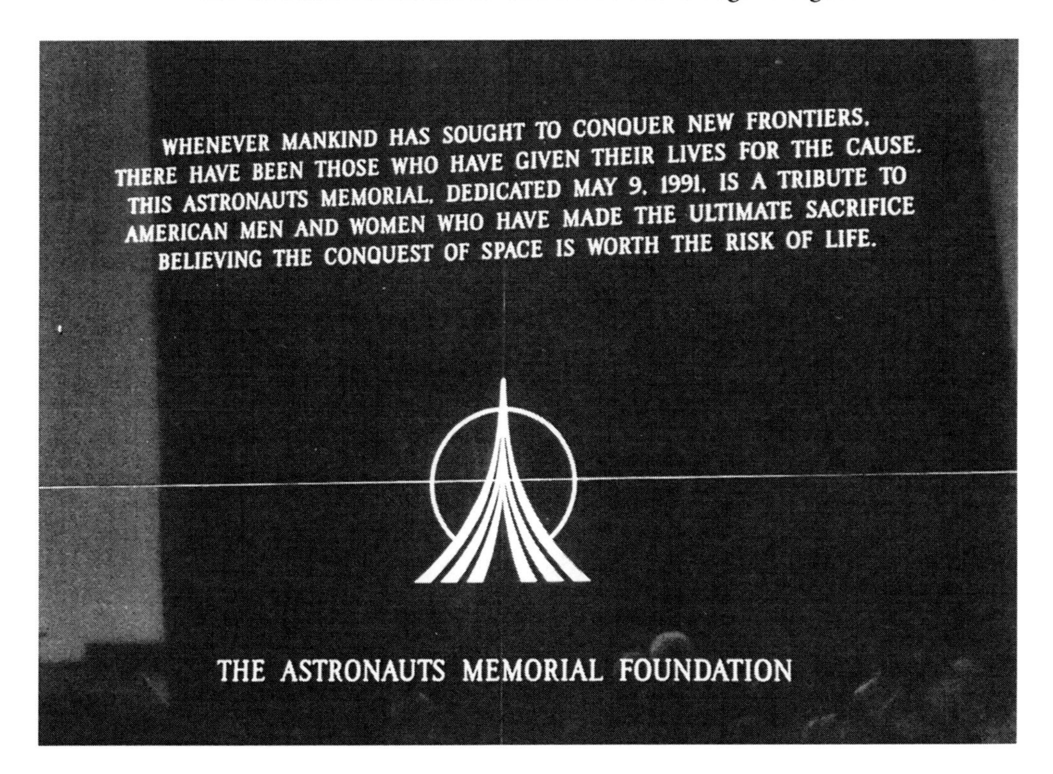

The inscriptions at the top of the Astronauts Memorial at Kennedy Space Center (author's collection).

In 1995, the only names inscribed on the Astronauts Memorial at Kennedy Space Center were those of the three astronauts who perished in the Apollo 1 fire (author's collection).

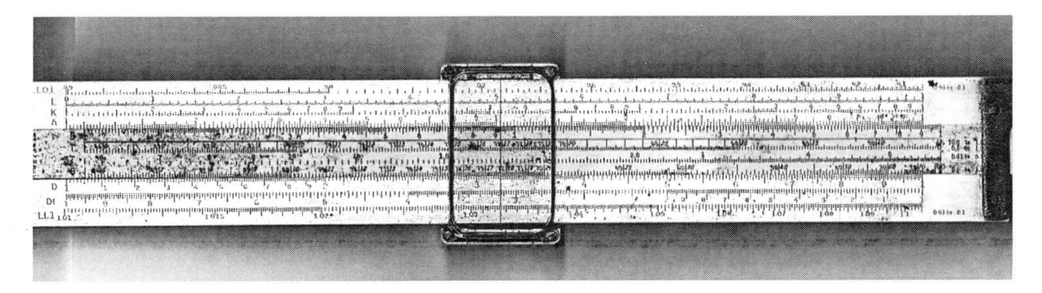

The slide rule that engineers used to do all the calculations required to design the Apollo Capsules in the 1960s (author's collection).

Engineer Vern Johnson of Honeywell regarding problems with the Rotation Hand Controller and Translation Hand Controller. During this time frame, I was talking to Bill De Viney, and he told me that he had attended a meeting with some NASA managers, and one manager told the attendees that the NAR Space Division had gone to the moon, all right, but they had gone about it all wrong. Bill's comment to the manager was "Yes, Doctor Salk, I know you discovered the vaccine, but you went about it all wrong." Meaning, of course, that success is success and that it is irrelevant and does not matter how you go about it.

Thursday, September 11, 1969, was my last day with the Apollo program; sometime that day, Carl Conrad told me that on Friday I was to report to Dave Engels at the Seal Beach facility to work on a special program. All night long I worried and wondered what type of program Carl was sending me to work on. What exactly was going to be required of me, and when was I coming back to work on Apollo? The Hispanic engineers I left behind in Carl's unit were Paul Garcia, George Cortes, and Pete Ontiveros. The remaining tasks for the unit were the launches of Apollo 12 through 17 and collaborating with the Russians in missions to an international space station. As far as I know, all three men worked on those tasks because I got to work with them again a year or so later.

The following day, Friday, September 12, I went to Seal Beach to work for Dave Engels, and after talking to him for a while, I learned that the program I was joining was the study contract from NASA named Integrated Launch and Reentry Vehicle (it was normally referred to as the ILRV contract), which was basically a feasibility study for a Space Shuttle. NASA had come to NAR Space Division and other companies and said, "We want you to design (to basically invent) a space truck to take payloads into orbit. We want it to take off like a rocket and land like an airplane. When it lands after delivering a payload to orbit, we want to just kick the tires to make sure they are okay, load another payload into it and launch it back into orbit." The comment the division managers made to themselves was "They're crazy."

NASA invented the idea of the Space Shuttle, but North American Rockwell in Downey created what had never previously existed. The designing and building of rockets and spaceships are not rocket science but the domain of the engineer. As the famous Dr. Theodore von Kármán put it, "The scientist describes what is, the engineer creates what never was."[10]

Epilogue

I first want to explain why I have dedicated this book to my late wife, who passed away in September 2014 due to Alzheimer's disease. I was her caregiver for at least the last ten years of her life, but that is not what I want to write about. Rather, I wish to tell readers about how she was a terrific wife who really made our married life an awesome one and helped me pave the way to a successful engineering career, both beyond my wildest dreams, and then I will say a few words about some of the engineers I was privileged to work with during my career up to and including the Apollo program.

I met my wife in high school in 1946, when she was a freshman and I was a sophomore. At that time, my two major life goals were to earn a college degree and to marry the girl I loved. Our love for each other withstood the test of time, which encompassed the three years of high school and our subsequent separation while I attended UCLA for one full semester and then served four years in the Air Force during the Korean War. We were married five months after I returned home following my discharge from the Air Force. In February 1956, we moved into a house we had bought, on a lot that was joined back-to-back with her older sister's house in Santa Fe Springs, California.

It took me three years at East Los Angeles College and three years at USC to earn my engineering degree while I also worked a full-time job. For all those years, my wife had my clothes ready every day; my breakfast, lunch, and dinner were always ready for me in time for me to get to my classes and my job; and the house was clean and well organized. She would always wait up for me when I returned home from work around midnight, and we would talk about our day.

She never complained even when I had to shut myself up in one of the bedrooms all weekend long to do homework and study for exams. She only complained whenever we could not attend a party or a dance because I needed to study. When I was promoted to a test equipment design engineer at Hughes, it became easier for us because then I was able to work a split shift, starting work right after my last class at USC and working part of the day shift and part of the night shift, which got me home earlier in the evening. I completed my studies for my engineering degree in the summer of 1960, and then we were able to live a normal life for six years before we moved to Minneapolis. She was not only my wife but also my best friend.

After I completed my last class, USC mailed me my degree. My wife knew exactly what we had accomplished by when I received my engineering degree. I did not attend my USC graduation with the class of 1961 because that is when I was in Fort Walton Beach, Florida, from March through most of May while working on the Hound Dog cruise missiles. The job assignment in Florida helped launch my successful engineering career, which resulted in my first promotion, and I owe it all to my wife for having agreed for me to volunteer for the assignment and agreeing to go with me.

She was very sympathetic to all my various attitudes and problems during the years I worked on the unmanned Apollo capsules. When I was offered a promotion that came with the requirement to move to Minneapolis to work on the manned Apollo capsules, she agreed to relocate and even sell our house, knowing full well that she would no longer be living next door to her favorite sister. When we returned to California, we both worked as a team to resettle our family, and during my work on Apollo 7, 8, 9, 10 and 11, she took it all in stride like a good trooper.

Even at this late date, I cannot forget Ray J. Hagerty, our department head at Hughes Aircraft Company, a tremendous engineer and a superb department head. He had faith in me and promoted me into the engineering department during my junior year at USC and taught me everything I know about digital computer theory, digital computer logic and binary arithmetic. Two other engineers I will forever remember are John J. Case and Dick Pieper, whom I met when I worked for Hughes Aircraft; Dick gave me a bird's-eye view of what it was like to be an engineer, helping to guide me in the right direction, and John gave me the opportunity to work the night shift so that I could begin my college engineering education. Ray F. Burke will always have a place in my heart because we went out to dinner every Friday evening while we both worked at Hughes. Ray took a leave of absence from Hughes to complete his engineering degree at the same time I quit Hughes and went to work for North American Aviation; later he received his master's degree from USC and eventually became a department head at Hughes.

From my involvement with the YITM group at North American Rockwell, the persons who stand out in my mind are Henry Pacheco, Monte Perez, Manny Alvarez, Tony Vidana, Jake Alarid, Hector Maldonado, Frank Serna, Hank Martinez, Phil Padilla, Gil Garcia, Fred Rodriguez, Larry Luera, Joe Gomez, and Ralph Gomez. I will always remember Henry Pacheco and Monte Perez because they gave of their time in coming to our division and explaining their participation in the East Los Angeles school walkouts and making us aware of the Hispanic high school dropout rate. Manny Alvarez and Tony Vidana were, of course, instrumental in getting us together so we could get organized to address our participation in helping to reduce the dropout rate. The other friends I mentioned above were very much involved with me in the YITM group and will forever be part of my life because we spent many hours together during our work in seeking to improve equality for minorities at Rockwell by making changes to company culture.

While I worked at North American Aviation Missile Division in Downey, California, there were several individual engineers I worked with who stand out in my mind. Among these, Paul Garcia, George Cortes, Bob Antletz, Claire Harshbarger, Ed Kelley, and Ken Watson were six phenomenal engineers. Paul Garcia was one of a few Hispanic engineers with a long tenure at North American Aviation Missile Division; he was a test engineer who was well respected for his expertise and had processed many Hound Dog cruise missiles through the staging area in preparation for launch by our test team in Florida, which culminated in our company selling 500 GAM-77 and 300 GAM-77A cruise missiles to the Air Force. (I will have more to say about Paul when I complete the book I am writing about the invention and design of the Space Shuttle.) I will forever remember Paul because we always worked well together, especially on the Space Shuttle program; unfortunately, I lost track of Paul in 1980 and have had no luck trying to locate him. George Cortes had expertise in some extremely critical areas of the Apollo and Space Shuttle systems and was well respected by everyone. During the Space Shuttle program in the mid–1970s, he was promoted to supervisor—a well-deserved advance; I lost track of George in 1980. Nor can I ever forget Bob Antletz, the man who had faith in me in 1960 and hired me to work on cruise missiles while I was still attending USC and helped me launch an unforgettable engineering career. I am sorry to report that I learned from his wife on a 2009 telephone call that Bob had passed away from brain cancer in 1997. Ed Kelley was the best mentor a person could ever wish for; he made me the engineer that I ultimately became. In February 1984, while I was working as a consultant for the Aerospace Corporation on an assignment in Sunnyvale, California, I was told by friends at Rockwell that Ed Kelley had passed away of a massive heart attack. I still think a lot about all the engineers I knew during the seventeen years I worked for North American Rockwell, and it is a shame to see all that talent disappear one by one.

A giant in my eyes while I worked on the unmanned Apollo capsules was our director, Dave Levine, because he was the one who told me that I owned Autonetics, giving me the green light to spread my wings. Years later, Dave and I came together again when I worked on the design of the Space Shuttle. Larry Hogan I admired because even though he was not my supervisor, he was a good supervisor to his engineers and was very involved in the design of the Control Programmer for AFRM 009 and the Mission Control Programmer for AFRMs 011, 017, and 020. When I was solving problems on the Rotation Controllers that were installed on two of the astronaut couches in the Apollo capsules, I needed to go into some of the Apollo capsules to gather some data so I could solve the problems. At the time, Larry Hogan was systems engineer on the Stabilization and Control System, and Larry helped me many times by calling someone in Building 290 and arranging for me to have access to an Apollo capsule when no one was in it so I could do my work. The last time I saw Larry Hogan was around 1982 when I was a project engineer at the Aerospace Corporation in El Segundo, California, at the NASA Manned Spacecraft Center. I was in one of the center's buildings waiting for an elevator when Larry approached the same elevator and we shared the elevator to an upper floor and had a good conversation.

Another person I must mention here is Bill Fouts, the best manager I worked for during my total engineering career, and I hold him in high regard although he never promoted me to anything (which might have been due to company culture at the time). I thank Bill for not holding me back during my work on the unmanned Apollo capsules and allowing me to bring the design and missions of the AFRM 009 Control Programmer and the Mission Control Programmer for AFRMs 011, 017 and 020 to a successful conclusion, which did so much for me in all my future endeavors. Thom Brown I will never forget; Thom was the engineer I took with me to Autonetics when he had just transferred from field engineering. He subsequently made a smooth transition to research and development work, doing well on the Apollo program and in the Space Shuttle program. All that made me very proud of Thom.

Every promotion I got in the years I worked for North American Rockwell came about through my mentor Ed Kelley. As I think about it now, I am thankful to God that I was not promoted to management levels at North American Rockwell because if I had become part of management, I would never have been involved with YITM and may never have left the company permanently, as I did in July 1980. That departure opened so many doors in my career that from 1983 to 2002 my wife and I traveled the United States and Europe on my jobs while I commanded a six-figure salary as a consultant or an independent contractor for various aerospace companies.

In closing these reminiscences, I can say that if I live to be one hundred years old, I will never forget the things and the people I have included in every chapter of this book.

Appendix

Notes Taken During My Support of Apollo 10 Mission—May 1969

APOLLO 10 (CSM 106)
MISSION SUPPORT LOG
J. De La Rosa
0.0 to 0800 PDT

May 19, 1969

0026 hours

Request by Flight to go over the day's plan & any updates on flight plans by 20:00 GET [ground elapsed time]. *(This is a request from the Flight Control NASA Controller to be allowed to review the day's plan.)*

0040 hours

40.32 ft/sec *(This lone entry must be a required added velocity—known as delta-V [ΔV]—because it is a mere 27.48 miles per hour and too small to be a velocity of the Apollo vehicle.)*

0045 hours

G & C Review *(Guidance and Control review)*

PIPA looking good. *(PIPA stands for Pulsed Integrating Pendulous Accelerometer, which was an element of the Inertial Navigation Platform that contained moving parts and tended to drift away from a calibrated position.)*

Platform drift—low. *(Platform refers to the Inertial Navigation Platform.)*

0.02 deg/hr highest drift
Update value 0.075 deg/Hr

G & N & SCS working good. *(Guidance and Navigation and Stabilization and Control System were working well.)*

SM RCS *(Service Module Reaction Control System)*

All Quads looking good—cooling good. *(Reaction jets in the Service Module, clustered in fours [quads], were all looking good and not overheating.)*

20 degree dead band—got jet firings. Crew is being awakened by the jet firings.

Proposed to keep from firing jets & waking up crew during passive thermal control mode (PTC) during Lunar coast. *(PTC was a passive thermal control mode whereby the Service Module, Command Module, and Lunar Module combination was put in a slow spin mode in the roll axis "a la barbeque style" in order to distribute the sun's heat evenly throughout the spaceship to keep it from overheating in any one section while the spaceship was coasting toward the moon. The spacecraft would normally wobble somewhat in pitch and in yaw and was allowed to do so only for a few limited degrees; when the limit was exceeded, the proper jets would fire to bring the spacecraft back in line. The limited degrees and the rate at which the spacecraft was allowed to wobble was called the dead band. In this case, the set dead band of 20 degrees was causing the jets to fire too often and waking up the crew. There was a proposal to make a change that could possibly solve the problem, written directly below.)*

Go from 20 degrees dead band attitude & 0.1 deg/sec dead band to 30 degrees dead band & 0.3 deg/sec dead band. *(The proposal was to widen the attitude dead band from 20 degrees to 30 degrees and increase the allowable rate from 0.1 degrees per second to 0.3 degrees per second.)*

[I] Got [a] call from Bob Hartley. [He asked me] What does Downey think of the proposed plan?

1. Maneuver to PTC attitude with CMC in minimum dead band. *(CMC is Ramon Alonso's Command Module Computer. The Command Module was maneuvered so that the side faced the sun with the Command Module Computer set to the minimum dead band.)*
2. Leave CMC in 0.5 deg dead band for five minutes. *(We would leave the Command Module Computer in 0.5-degree minimum dead band for five minutes in order to allow the spacecraft to establish the proper passive thermal control attitude without wobbling.)*
3. Spin to 0.3 deg/sec and disable roll jets. *(This action would set up the spacecraft's passive thermal control spin speed with the use of the roll jets. The roll jets would then be disabled to prevent them from inadvertently firing and disturbing the passive thermal control spin speed.)*
4. Leave Pitch and Yaw in minimum dead band for five minutes. *(This would ensure that the spacecraft was not wobbling.)*
5. Establish 30 degree dead band. *(After five minutes, the Command Module Computer would be set to the 30-degree dead band.)*

[I] Told him it looked OK but that I would check with our people. *(It seemed reasonable to me that changing to a wider dead band would reduce the yaw and pitch jet firings.)*

0117 hours

[I] Called Al Sohler & got him out of sack in Houston. [He] Said procedure was OK for elimination of jet firings. *(I went through the whole procedure with Al Sohler because he was in Carl Conrad's unit and was responsible for the guidance and control of Apollo 10, which is why he was in Houston at the NASA Manned Spacecraft Center.)*

[I] Called Bob Hartley back & told him OK.

0125 hours

Flight Plan

- No major changes.
- Wake up crew at 21:30 GET per plan.

- P23—New stars to be selected. LM in way—cannot use stars in flight plan *(LM means "Lunar Module." Here the chosen star to be viewed for navigation could not be seen because the Lunar Module was obstructing the view, so another star had to be selected.)*
- Will probably experiment with new methods (deviate from flight plan) on PTC before nite is over & thus have not made definite changes to flight plan. Change will probably be made later.

0143 hours

Yaw jet firings reported.

Indications: All gyros off (3 BMAGS). *(The three body-mounted attitude gyros in the Stabilization and Control System that were the backup attitude reference system for the Primary Guidance and Navigation System were turned off.)*

Chart recorder Cal—wipes out data. *(The chart recorder calibration would wipe out all data.)*

State Vector
78 h 58 m
1680 nm
Vrho = 12.58 n
66 degrees E
} *This is all navigation data.*

20 degrees off from x-axis—rotation
PTC—How about maybe go to drift in roll & locate axis.
} *Questions about the x-axis location during passive thermal control mode.*

0415 hours

PTC *(passive thermal control)*

Rate of consumption during PTC—fuel? *(The big worry was that the jet firings during passive thermal control mode would deplete the onboard reaction jet fuel.)*

3 hours from now when crew wakes up the consumption S/B almost the same (G & C Houston computing). *(S/B stands for "should be.")*

CONSUMPTION

S/B 1.0 Lbs/Hr (Spin and to hold PTC)

Estimate at 1.5 Lbs/Hr (Spin and to hold PTC)

The above numbers do not include fuel used to maneuver to PTC attitude.

CDR—sleeping OK during PTC mode. *(CDR is Commander Thom Stafford.)*

CMP—Having trouble sleeping during PTC jet firing. *(CMP is Command Module pilot John Young.)*

0437 hours

Range & range rate of S-IVB relative to CSM/LM. *(Location and speed of the Saturn IVB booster that was ejected earlier relative to the Command-Service Module/Lunar Module spacecraft.)*

R = 1100 nm
R rate = 136 ft/sec
} *1,100 nautical miles and a speed of 136 feet per second*

IN 10 HOURS

1954 nm
R rate = 154 ft/sec
This was where the Saturn IVB Booster would be in 10 hours relative to the spacecraft.

S-IVB trailing CSM/LM behind, below and to the left of CSM/LM in curvilinear system.

0620 hours

[I] Called Jack Jensen and gave him PTC problem & recommendation. He will call Ben Lum. *(Jack Jensen and Ben Lum were two other experts in our Guidance, Navigation and Control Department.)*

0710 hours

Sohler Telecon
Scuttle ??? something??
Yaw or pitch jet firing every 5 min.
Data will come late this afternoon.
This was a telephone call I had with Al Sohler, and it seems that someone wanted to scuttle something because the yaw or pitch jets were firing every five minutes.

0724 hours

49 ft/sec MCC planned. *(MCC is "mid-course correction." A 49-feet-per-second delta velocity mid-course correction was planned by firing the Service Propulsion System engine.)*

0725 hours

Flight Plan Change

- Stars to be corrected. *(Star sightings to be corrected because the Lunar Module blocked the view of the planned stars.)*
- MCC2 49 ft/sec. *(This would be mid-course correction number two on the way to the moon.)*
- PTC problem being worked. *(The excessive jet firing problem during passive thermal control mode was being investigated.)*

0726 hours

CSM/LM Contacted.

A—*(I have no data recorded.)*
B—94.5
C—96 %
D—92 %
This appears to be the fuel remaining in each of the Service Module Reaction Control System jet quads from A to D.

Updated Flight Plan.
CDR
"Noise & oscillations the days out." *(Thom Stafford was complaining.)*
No big deal.
Gene slept comfortable.

0745 hours

Stafford comment:

"Everybody took a drink of water, and it was good, but when I took a drink, it was lousy. My mouth is still burning. Just wanted this on record."

0748 hours

Al Sohler Telecon: *(Notes I took during a telephone conversation I had with Al Sohler in Houston.)*

Thermo, Gimbal, lock, tumbling, OK. *(It seems like Al was saying that a gimbal lock and a gimbal tumbling of the inertial measurement unit was not a problem.)*

GET 2400 meeting. *(I believe that Al was saying that either the above was okayed at a 2400 hour elapsed time meeting or that he was going to attend a 2400 hour elapsed time meeting.)*

May 20, 1969

00:00

Reporting in to relieve Paul Garcia.

0029 hours

Trajectory status. *(Status of the trajectory to the moon.)*

Post MCC2 *(The second mid-course correction had already been executed.)*

Pericynthion *(The normal orbit of one body around another body is elliptical. For example, when a satellite is placed in an orbit around the Earth, it normally assumes an elliptical orbit with the Earth at one foci of the ellipse. The point of the elliptical orbit where the satellite comes closest to Earth is called the perigee, and the point where it is farthest away from Earth is called the apogee. When Apollo 10 went into moon orbit, it assumed an elliptical orbit in which the point of the elliptical orbit where the spacecraft came closest to the moon was called the pericynthion, while the point of the elliptical orbit where it was farthest away from moon was called the apocynthion.)*

66.5 nm
58 nm
55 nm
61.5 nm—Steady now with 10 Hrs of data.

} *These are the pericynthion values of the various elliptical moon orbits being considered for Apollo 10 when it reached the moon. 61.5 naut. miles seemed the best.*

LOI—Channel 23 solution looks good. *(The channel 23 solution for lunar orbit insertion [LOI] looked good—whatever channel 23 was.)*

LOI1 = 2970 ft/sec *(This appears to be a delta velocity—that is, a change in velocity—for lunar orbit insertion because 2,970 feet/second converts to 2,025 miles per hour, which is too low for total spacecraft velocity.)*

0037 hours

Battery charges working good.

Hi Gain Antenna didn't break lock.

0430 hours

MCC3 might not be necessary. *(Mid-course correction number 3 might not be necessary to ensure that the trajectory to the moon was correct.)*

- Thruster config? *(The question as to what the Service Module Reaction Control System configuration will be.)*

Well balanced—Can go to any config. *(The Reaction Control System could go to any configuration because the quads were well balanced.)*

G & C would like to consider this for a minute. *(Guidance & Control Ground Controller would like to consider this option.)*

0436 hours

A C roll to be used exclusively today for SCS or G & N. *(The A and C roll quads were to be used on this day exclusively for Stabilization and Control System or Guidance and Navigation.)*

0523 hours

Established communication with crew. Will let them get up at their leisure.

0532 hours

Crew listened to update news while preparing breakfast.

154,423 nm out. *(Apollo 10 had traveled this far in nautical miles from Earth.)*

0539 hours

Gene Cernan

Reports Africa is clear & countries Northeast also.

May 21, 1969

0047 hours

LOI to slip about 10 min. *(Lunar orbit insertion had slipped about 10 minutes.)*

0115 hours

For PTC Mode—Do step 1 of DAP is already active. *(DAP was the digital autopilot, which was an application software—an app—located in the Command Module Computer that flew the Apollo spacecraft to the moon and back.)*

0147 hours

MCC4—Looks like this will not be done. Crew will sleep longer if MCC4 not made. *(It looked like the fourth mid-course correction would not be done, which meant that the Service Propulsion System Engine would not be fired.)*

LOI time slipped 10 min.

This slips FLT. PLAN by 10 min. *(The slippage of the lunar orbit insertion time has delayed the flight plan by 10 minutes.)*

PTC

Excursion more erratic this last nite. Could possibly fire a jet before nite is over.

0225 hours

SCS is GO. *(The Stabilization and Control System was ready to go.)*

PTC

Two schools of thought

1. Active by 28 degrees in pitch. Later will switch to yaw and no jets will fire.
2. 69–70 hours GET jets will fire.

0245 hours

PTC

Procedure for various ways of starting PTC.

Don't use verb 48 or 46. *(Verbs were CMC entries implemented via the display key entry.)*

Current Traj. *(The current trajectory below.)*

Approx. 61 nm

Pericynthion = 61.14 nm

76:10 GET Arrival time.

3.5 ft/sec MCC4 correction.

0400 hours

Vsub-x = 0.7 ft/sec

LOI—would like 60 × 70. Target by 60—this has 2 mi uncertainty could be 62 & could not then be circularized to 60. Considering targeting for 58 LOI and with uncertainty could end up at 60 or 56 but could still circularize at 60. *(When the Apollo 10 Service Propulsion System engine was fired for lunar orbit insertion, the spacecraft, on being captured by the moon's gravity, would normally go into an elliptical orbit. However, a circular orbit was desired, which was accomplished by firing the Service Propulsion System engine at apocynthion or pericynthion after the elliptical orbit was established. So, the planned elliptical orbit had to have the correct apocynthion and pericynthion for the desired circular orbit. In the case of Apollo 10, it needed to be placed in a 60-mile circular orbit above the moon. At this point in the mission, it looked as though placing the spacecraft in an elliptical orbit 60 × 70 would not do the job because, with the uncertainty of 2 miles, it could end up with a pericynthion of 58 miles, which could be circularized to 60 miles or 62 miles, but this could not be circularized to 60 miles. It would be better to target for a pericynthion of 58 miles, which, with 2-mile uncertainty, could end up at 56- or 60-mile pericynthion, both of which could be circularized to 60 miles.)*

0417 hours

Talking about wake-up time of 70:00 GET.

0422 hours

FLT—FIDO *(Flight Fido is the code word for the NASA flight control system controller.)*

Committed to no MCC4 at this time.

0535 hours

No MCC4 affirmed.

0545 hours

FLT—FIDO

Brief all controllers

1. No MCC4.
2. Let crew sleep.

A C Roll stay with this quad config. *(A and C jet quads in the Service Module would be used for the roll, so we needed to stick with this configuration.)*

Pericynthion = 61.1 nm

Arrival time = 76:0:10

LOI—11 min late.

MCC1 to MCC2 change caused 11 min difference.

0622 hours

61.9 pericynthion.
76:00:12 arrival time.

0640 hours

27.7 deg Pitch PTC *(Spacecraft pitch attitude of 27.7 degrees for passive thermal control mode.)*

Should not go over 29 deg.
Peak-to-peak 3 Hr period.
26.7 deg Yaw PTC. *(Spacecraft yaw attitude for passive thermal control mode.)*
Both Pitch & Yaw should peak at 29 degrees.

0702 hours

LOI *(What follows are data for a lunar orbit insertion.)*

75:55:54 *(This insertion would occur at this point in ground elapsed time.)*

18 degrees *(This would be the entry angle.)*

LPO—Long 156 degrees E *(LPO is lunar parking orbit. This parking orbit would be at a longitude of 156 degrees east.)*

98 mi Alt. of burn ignition. *(This would be the altitude from the moon at ignition of the Service Propulsion System engine for orbit insertion.)*

Gamma at burn out. *(Gamma was the angle between the Apollo fight path and the moon horizon at Service Propulsion System engine burnout.)*

-0.88 degrees *(This is the Gamma angle.)*

Alt. at burnout 69.1 nm *(For this lunar orbit insertion, the Service Propulsion System engine burnout would occur at 69.1 nautical miles altitude from the moon.)*

Δangle = 5.5 degrees (plane change). *(This step would require a delta angle plane change of 5.5 degrees, possibly to be able to go over candidate landing sites for Apollo 11.)*

59.5 × 170 *(The orbit would have a pericynthion of 59.5 nautical miles and an apocynthion of 170 nautical miles.)*

ΔV = 2982.3 ft/sec
5 min 54 sec burn.
ΔV range after burn = 2597 Docked
Twice for undocked

May 22, 1969

0020 hours

Houston check communications to Honeysuckle okay. *(The communications between Mission Control in Houston and the Honeysuckle Creek radar tracking station in Australia were confirmed as being established.)*

0031 hours

Cap Com contacts Apollo 10 crew. *(The only person allowed to talk to the crew during the whole mission was one of the astronauts, whose call name for this purpose was Cap Com, an abbreviation for "Capsule Communicator.")*

The crew was asked to turn off one each pitch, roll and yaw [attitude] reaction control jet.

The B and C reaction control jet quads are all off. *(Verified that the reaction jets requested to be turned off had been turned off.)*

0036 hours

Crew was told to go to sleep and to expect more reaction control jet firings because the spacecraft is in the hold attitude mode. *(When the spacecraft was in the attitude hold mode, it was put in a selected attitude while the crew slept, and then if the spacecraft strayed a specified number of degrees from the selected attitude, the reaction jets would fire to bring the spacecraft back in line. This was done so that the spacecraft communication antennae were facing in the optimum direction so that the ground could monitor spacecraft activity and be able to communicate with the crew to wake them up.)*

0040 hours

The Lunar Module Pilot is in his sleep station. *(Eugene E. Cernan)*

The Commander is asleep. *(Thomas P. Stafford)*

The Command Module Pilot is awake but going to sleep. *(John W. Young)*

The pitch jets are firing on the average of every 5 seconds. The hang up is due to only one jet authority for such a big inertia. *(Inertia is the resistance of any physical object to a change in motion when at rest or in motion. The "big inertia" was a reference to the Command Module–Service Module combination that was in an attitude hold mode; when it moved away from the attitude it was supposed to be holding, the reaction control jet would fire to bring the spacecraft back to the correct attitude. The problem was that there was only one jet, consisting of a quad, for each of the two directions in yaw and the two directions in pitch. The lack of any other jets meant that this small jet had to fire more than once to reverse the direction of such a big spacecraft.)*

0110 hours

The reaction control jet selected to fire to control the command module pitch axis is the A quad.

The reaction control jets selected to fire to control the command module roll axis are the A and D quads.

0118 hours

The Flight Surgeon says that it looks like both crewmen are asleep. They are probably not being awakened by the reaction control jet firings.

0225 hours

The Mission Control Guidance and Control System Engineer reports that the command module is in the middle of the dead band. Looks okay now, not hanging up.

0230 hours

The Mission Control Guidance and Control System Engineer reports: The command module is approaching the dead band in pitch.

0234 hours

The command module yaw axis is now turning around. Estimate is that it takes 3 to 5 firings of jets to turn the command module around.

The Mission Control Guidance and Control System Engineer reports: Looks like we will be okay now.

0235 hours

Loss of signal (LOS) occurs at this time. *(When the orbiting Command Module went behind the moon, the communication signal with the Command Module was temporarily lost.)*

0533 hours

The plan is to wake up the crew halfway through this trip.

AOS on next trip at 9230 GET. *(AOS is acquisition of signal.)*

Flight Surgeon says: The Command Module Pilot is awakening. *(John Young)*

0612 hours

7 minutes from acquisition of signal (AOS). *(At this point, we were 7 minutes away from the Command Module coming out from the back side of the moon, allowing us to be able to communicate with them.)*

SPS

Believe PUGS

(At this time, I have no idea what the note with PUGS means except that the Service Propulsion System [SPS] engineer was giving us some information.)

314 lbs—reduce S/C wt. *(This note says that the spacecraft needed to be reduced in weight by 314 pounds.)*

500 # reduced usable oxidizer & fuel.

(The above note no doubt indicates that the Command Module had reduced the usable oxidizer and fuel by 500 pounds.)

ΔV capability 4700 ft/sec

REV 10

(The above notes from the SPS engineer tell us that the Service Propulsion System on the Command Module at this time was capable of imparting an added velocity of 4,700 feet per second. I don't understand the note on the revolution number 10 around the moon because that had occurred much earlier.)

0619 hours

The Flight Surgeon says he will advise when the crew is awake.

0625 hours

Flight Surgeon says sleeping very lightly and could wake crew up if desired.

0632 hours

Quad A CSM

118

101

97 He

(The above note probably gives the status of the pressure of the fuel, oxidizer and helium of the A quad Reaction Control System jet arrangement on the Command-Service Module. The status was provided by the NASA engineer sitting at the Command-Service Module fuel and oxidizer console in Mission Control in Houston.)

Must roll S/C 180 degrees because of thermo for one hour.

Quad Alpha He will go into Redline.

(The two notes above recommend that the spacecraft be rotated 180 degrees because of the sun heating up the helium; if the spacecraft was not rotated to cool off the A [Alpha] quad helium, the helium would go into the redline region, putting the whole spacecraft and crew in danger.)

Surgeon says crew stirring around.

What does 180 roll do to communications?

—Lose Hi Gain Antenna. *(No communication with the crew via the High Gain Antenna.)*

—Must go to Omni Antenna. *(Must start communicating with the crew via the Omni Antenna.)*

Waking up crew—WHAT MUSIC! FRANK SINATRA, "THE BEST IS YET TO COME."

(Waking up the crew was always a big deal. As part of awakening the astronauts, Mission Control would play some great music for them. This time they played the Frank Sinatra song "The Best Is Yet to Come.")

0648 hours

The news was read to the crew and also their horoscopes were read to them.

0654 hours

Roll 180 degrees at LOS & Roll back to proper attitude at AOS. *(This note says that the Apollo Command Module would rotate 180 degrees at loss of signal when it went behind the moon and would rotate back to its proper attitude at acquisition of signal when the Command Module reemerged.)*

0656 hours

Expect LOS at 9342 (GET) *(Loss of signal expected at 9342 ground elapsed time.)*

AOS at 9429 (GET) *(Acquisition of signal expected at 9429 ground elapsed time.)*

Roll Δ of 180 *(This note just says that the Command Module would rotate 180 degrees.)*

0700 hours

6 HRS. sleep *(The crew has had six hours of sleep.)*

Crew Report *(Crew report given by Gene Cernan.)*

(Gene)

Gene Cernan reports three equations, one for Roll Rate, one for Pitch Rate and one for Yaw Rate.

0720 hours

(Someone reported the sunrises and sunsets that occurred on Revs 10 and 11.)

Rev 10

093 52 22 Sun Rise

095 04 46 Sun Set

Rev 11

095 41 06

095 52 52

096 27 16

Sun Rise

095 50 58
Sun Set
097 03 22

0726 hours

Clutch Diff. Current?
Al Sohler
Al Sohler and Bob Field looked at post LOI [lunar orbit insertion] Clutch current data
Avg Diff Current
Yaw = 89 milliamps
Pitch = 79 milliamps
Max Yaw = 115 milliamps
Max Pitch =100 milliamps
Not uniformly ablating
No Problem
DB [dead band] problem during LO [lunar orbit] rest cycle
Looks like due to gravity gradient
.01 to .04 sit & limit cycle in yaw then switched to pitch & same for pitch & then switched back to yaw
59-minute cycle
LOI 1?
Hot engine could expect this type of thing

May 23, 1969

00:00 hours

CAP COM is communicating with the crew. *(Apollo 10 was in moon orbit at this time.)*

CAP COM is telling the crew about the new wakeup time. The new wakeup time will be 121 hours which is acquisition of signal (AOS) on revolution 22. *(AOS occurred when the Apollo Command Module came out from behind the moon during an orbit and communications from Earth could be reestablished with the crew.)*

Crew wants to know where they stow stuff they brought back from "Snoopy." "Snoopy" is 6000 miles [from] "Charlie" and has 120 hours of power left. *(Snoopy was the name of the Lunar Module, which had been jettisoned the day before after completing its mission of a partial moon descent; Charlie Brown was the name of the Command Module.)*

0030 hours

Revolution 18 has had only one jet firing.

0055 hours

The Command Module is in the wide dead band mode (+/- 5-degree dead band about a diagonal axis). *(A dead band for the Apollo Command Module described the degree to which the attitude of the Command Module would be allowed to stray. If there were no limits to the control of the Command Module attitude, the command module could begin to tumble, which was not desirable. A +/- 5-degree dead band allowed the attitude of the Command Module to stray +/- 5 degrees in attitude before reaction jets on the Service Module would begin firing to bring the command module attitude back to its zero-degree point.)*

The Command Module attitude hold is now in minimum dead band mode. *(This was a tighter dead band that did not allow the Command Module to stray far from a present attitude, thereby allowing the Service Module reaction jets to fire more often and use up more valuable fuel.)*

0140 hours

The Command Module attitude control mode is now going to max dead band mode for attitude control.

0142 hours

Max dead band affirmed now.

0252 hours

TEI *(Trans-Earth injection)*
Over 4000 feet per second delta velocity left

0317 hours

4754 feet per second delta velocity capacity

0350 hours

Tweeked RCS for 20 seconds
Worked like charm

0353 hours

Madrid & Honeysuckle check for communication

0420 hours

0.2 degrees per second rate

0449 hours

Crew to sleep late & delete rest period in middle of plan—puts TEI at same time in plan.

0505 hours

Depleted System A

0548 hours

SCS [Stabilization and Control System] working good

0607 hours

Seems like John Young is awake

0612 hours

Seems like crew is stirring around.

0622 hours

The crew is up.

Apollo 10 & Houston communications *(This indicates that the crew and Cap Com in Houston were communicating.)*

0625 hours

46 minutes to LOS *(loss of signal)*

G & C [Guidance and Control in Houston] recommends B & D roll jets

LM 23 K mi from moon *(The Lunar Module was ejected into space after completing its mission in approaching the moon and was at this point 23,000 miles from the moon.)*

0645 hours

Camera has faulty gear in back

0705 hours

Photo

0738 hours

56.3 × 55.5

May 24, 1969

0010 hours

Use pitch B & D *(use B and D Reaction Control System quads for pitch control)*

Horizon

6-degree window marker

Ignition—1 minute

Sextant star not available

Sun not available in COAS at ignition *(COAS, or crewman optical alignment sight, was the onboard telescope used by the crew to align the Command Module to the celestial body to be used for navigation sighting.)*

Horizon will be lit at ignition *(This note indicates that the horizon would be used for a navigation fix instead of the sun.)*

0028 hours

ORDEAL *(This is a mode used for the attitude displayed to the crew on the Fight Director Attitude Indicator while orbiting the moon.)*

Pitch 180

Roll 282

Yaw 000

20 degrees

T2 Yaw 20 degrees

High Gain—would like to have

TEI on next Revolution?

0035 hours

Telecon with Roy *(This note refers to a telephone conversation, probably with Roy Murphy.)*

Risky to throw out anything but if must then GDC, BMAGS and FDAI.

We can consider it later if necessary.

(I don't recall what this note is about, but I'm speculating that I placed a phone call to Roy to discuss the possibility of which Stabilization and Control System devices could be powered down in case of emergency to save electrical power.)

0040 hours

B & D quad for ullage. *(The B and D quad Reaction Control System engines in the Service Propulsion System were selected to fire and move the Command-Service Module combination in a direction to shift any residual fuel in the Service Propulsion System engines away from the combustion element to avoid an explosion when the engine was fired—this maneuver is referred to as ullage.)*

DAP loaded up. *(The Command Module's digital autopilot computer had all the trans-Earth injection and Earth entry navigation, guidance, and flight control data loaded into its memory.)*

Enable all auto RCS *(These are instructions to enable all automatic Reaction Control System jets.)*

A C for Roll—Affirmative *(Houston affirmed that the A and C Reaction Control System jets were to be selected for firing to control the capsule's roll maneuvers.)*

0046 hours

F C 2 &3 look good *(Fuel Cells 2 and 3 look good.)*

Looking at F C *(Houston confirmed that they were looking at the fuel cells.)*

0047 hours

F C 2 & 3 Temp in condenser being monitored 8–9 degrees spread. *(Fuel Cells 2 and 3 temperature in condensers were being monitored and showed an 8–9 degree spread.)*

40 degrees per hour going down

F C 1 Sw as function of time and not temp. *(Fuel Cell 1 would switch in as a function of time and not of temperature.)*

O2 flow rate up & down on FC 2 & 1. *(The oxygen flow rate was going up and down on Fuel Cells 2 and 1.)*

Flow meter not too good. *(Houston speculated that the problem was with the flow meter rather than the fuel cells.)*

0049 hours

6 min to LOS

Given them everything. *(Houston asserted that they had given the crew everything they needed to enter into the computer for trans-Earth injection.)*

149.5}Limit *(This temperature and the one below were the limits for fuel cells.)*

177 }Temp

154.2 *(This temperature and the one below were the temperature readings for Fuel Cell 2.)*

167

F C # 2

13–14 degree spread oscillations *(This is an affirmation by Houston that there was a difference of 13–14 degrees between the two readings of Fuel Cell 2.)*

0130 hours

TEI 32 PAD

Final TEI 31 PAD *(PAD is a project approval document containing all the required flight data for trans-Earth injection.)*

0135 hours

F C # 1 should come on the line.

0147 hours

PAD Update

(Notes taken in this time period were updated velocity data, attitude data and firing times for B and D quads, and I will not show it all here.)

0200 hours

(Notes taken in this time period were also loss of signal data and trans-Earth injection data, which I will not show.)

0225 hours

OMNI B Antenna *(The OMNI B antenna would be used for communication.)*

0227 hours

20 min to LOS *(20 minutes to loss of signal.)*

0455 hours

P23 might be deleted to let crew sleep 4 more hours.

TEI Burn 165.2 sec. *(The firing of the Service Propulsion System engine for trans-Earth injection would be for a duration of 165.2 seconds.)*

R 105

P 190

Y 0

(The above notes are the spacecraft roll, pitch and yaw attitude.)

(END OF MY NOTES)

Chapter Notes

Preface

1. Charles Fishman, "What You Didn't Know about the Apollo 11 Mission," *Smithsonian Magazine* (June 2019).

2. Around 2015 or 2016, when I was preparing my manuscript for publication, I Googled some Apollo-related websites for additional information. See Apollo Guidance Computer History Project, "Ramon Alonso's Introduction," updated December 8, 2002; Apollo Guidance Computer History Project, "Hugh Blair-Smith," January 2002.

3. I Googled Dr. Alonso's name and checked other sites, where I got a great deal of information about him (such as Hack the Moon's "Ramon Alonso, Designer of the DSKY, Computer Logic and Rope Memory"). If the reader wants more information on Dr. Alonso, just Google his name or "Designer of the Apollo Computer" for some interesting reading. For those readers who have older relatives who are Spanish speakers only, if you check enough sites, you can find Ramon Alonso articles in Spanish for them to read.

4. The list of names I listed in the second of these two paragraphs was received from Elisa Munoz, who did most of the typing needed for letters from YITM and whose name also appears in the list.

5. Much has been written about the Aztecs and their empire, especially about the Aztec Eagle Knight and the Aztec Jaguar Knight. One book in my possession that mentions the Eagle Knight is Gene Stuart's *The Mighty Aztecs*, published by the National Geographic Society in 1981, as well as a recent *National Geographic History* publication that includes an article titled "Aztec Age: Rise of an Empire," dated July/August 2021 (pages 60–74).

6. I had never heard of the Order of the Aztec Eagle, so I Googled it. Most of the information included here came from that web search. See Jen Kirby, "Jared Kushner Is Getting an Award from Mexico, and Mexicans Aren't Happy about It," *Vox*, November 28, 2018, https://www.vox.com/world/2018/11/28/18116287/jared-kushner-president-mexico-award-aztec-eagle-nieto-trump.

Introduction

1. Readers may already know that John F. Kennedy is considered a visionary because in his May 25, 1961, speech he set a goal for the United States to land on the moon by the end of the decade. However, if you want to know why the other three individuals are considered visionaries, you can find out by reading Ernst Stuhlinger and Frederick I. Ordway III's *Wernher von Braun: Crusader for Space* (Malabar, FL: Krieger, 1996).

2. R.H. Goddard, "A Method of Reaching Extreme Altitudes," Smithsonian Institution, 1919, https://transcription.si.edu/project/8542.

3. See Stuhlinger and Ordway, *Wernher von Braun: Crusader for Space*. This book includes some of the history of the space program and is a source of some of the space facts I have included in this book. If the reader wants to read about how the Apollo Program was created, go online and Google "Chariots for Apollo."

Chapter 1

1. Weapons of Mass Destruction Around the World, "SM-64 Navaho," July 17, 1998, https://nuke.fas.org/guide/usa/icbm/sm-64.htm. In 2000, I Googled "SM-64 Navaho" to verify what Ed Kelley had told me about the Navaho Program and the X-10 missile. The reader can Google that site and "Navaho Program" for more information and Navaho launch videos.

2. Ernst Stuhlinger and Frederick I. Ordway III, *Wernher von Braun: Crusader for Space* (Malabar, FL: Krieger, 1996). If readers want to learn more about the complexity of the problem, they are encouraged to read the details provided in this book. Readers can also Google "how did lunar orbit rendezvous get selected for Apollo."

3. NASA Electronics Research Center, "Apollo Command and Service Module Stabilization and Control System Design Survey: Design Criteria Program Stability, Guidance, and Control," December 20, 1968. The Block I/Block II Apollo capsule concept is discussed on page 5 of this booklet, which is in my possession.

4. There are two excellent websites where the reader can go to learn how important it was for the Apollo capsule to stay inside the reentry corridor.

One is "Working Out the Problems of Apollo 13," written by Candler Hobbs, which includes an interview with two Georgia Tech graduates, Spencer Gardner and Jack Knight, who were flight controllers during the ill-fated Apollo 13 mission; during the interview, they discuss the spacecraft's reentry. The other source is "Returning from Space Re-entry."

5. "Harrison Storms," https://en.wikipedia.org/wiki/Harrison_Storms. I had heard so much about our division president in 1964 that in 2005 I Googled his name, which is where I got the information about him that I included in this book. If the reader would like to know more about Stormy, Google "Harrison Storms aeronautical engineer"; there is a wealth of information available on his career.

Chapter 2

1. "AGM-28 Hound Dog," updated April 8, 2022, https://en.wikipedia.org/wiki/AGM-28_Hound_Dog. The Hound Dog Missile information that I have included here came from my experience working on that program, combined with information found online.

2. By the time I started writing this manuscript in 2000, the internet had become a valuable resource, so I went online to verify what Ed Kelley told me about the Navaho booster and the X-10 cruise missile. If anyone is interested in reading history on the Navaho missile, do a Google Search on "Navaho Missile Program"; there are multiple sites with Navaho history, complete with pictures and videos. One can start by checking out Wikipedia's article "SM-64 Navaho."

3. The development of the probe and drogue rendezvous technique during the flights of Gemini 6 and Gemini 7 is described in detail in North American Aviation's *Skyline Magazine* 24, no. 2 (1966).

Chapter 4

1. A lot of what I have written about Charlie Feltz circulated through the grapevine; I also knew some engineers and managers who worked for Charlie and had occasion to butt heads with some of them, which I will write about later.

2. This description of the boilerplates came from six issues of the *Skywriter* company newspaper dated December 11, 1964; December 18, 1964; December 31, 1964; March 26, 1965; May 21, 1965; and June 25, 1965. If the reader desires more details, Google "Apollo Program Boilerplate Capsules" for a wealth of information not only on Apollo but also on other space programs.

3. A discussion of the analysis and the resulting solution for the reaction control thrusters' solenoids is documented and accompanied by the circuit diagram of the design required to solve this problem in NASA Electronics Research Center, "Apollo Command and Service Module Stabilization and Control System Design Survey: Design Criteria Program Stability, Guidance, and Control," December 20, 1968, pp. 68–69.

Chapter 5

1. The astronauts' assignments and the Apollo Mockup review were reported in the July 17, 1964, issue of *Skywriter.*

Chapter 6

1. Bill Paxton, "Unit Organization—Automated Systems Design" (internal letter), July 29, 1964.

Chapter 7

1. *Skywriter* (North American Rockwell), January 8, 1965; *Skywriter* (North American Rockwell), January 15, 1965.

2. *Skywriter* (North American Rockwell), January 15, 1965.

Chapter 8

1. Let me clarify the six mandatory design changes made to the baseline Control Programmer and why these changes had to be made to the Production Control Programmer. Because John Rowe used the same schematic of the original mission design that he and I used to redline the mandatory design changes to incorporate the Rev A, Rev B and other approved modifications to come up with the production version of the Control Programmer schematic, that schematic also did not have the mandatory design changes to it, so those changes had to be added to the Production Control Programmer schematic; adding them would create further cost and time to the hardware, which had to be included in the firm cost proposal from Autonetics for the Production Control Programmer.

2. *Skywriter* (North American Rockwell), April 9, 1965, p. 1.

3. *Skywriter* (North American Rockwell), May 14, 1965, p. 2.

4. *Skywriter* (North American Rockwell), April 16, 1965.

Chapter 9

1. If the readers want to read these articles, they can search the *LA Times* archives online.

2. *Skywriter* (North American Rockwell), July 9, 1965.

3. *Ibid.*

4. *Skyline* 25, no. 3 (1967). The Apollo heat shield information was paraphrased from the aforementioned edition of *Skyline,* which was full of news about the Apollo Program. Other copies

of the *Skyline* magazine that are in my possession have articles with titles such as "Apollo Gets Ready for Moon Shot," "Tests Conducted All over Nation," "Engine Tests at Santa Susana, Landing Tests of Capsule at Mojave Desert," "Engine Tests at Mississippi Test Facility," "Ablative Heat Shield Tests at AVCO/RAP Corporation in Lowell, Massachusetts," "Water Landing Tests in Downey," "Dummy in Capsule Vibration Test in Downey," "Bellows for Navigation Telescope Tests at MIT/AC Spark Plug Labs," "S-V Bulk Head Built at Boeing in New Orleans," "Engine Tests at Marshall Space Flight Center in Huntsville, Alabama," "Service Module Proof Tests in NAA New Mexico Facility," "CM/SM Environmental Tests at Arnold Engineering in Tullahoma, Tennessee (Hot & Cold)," and "CM/SM Wind Tunnel Tests at Arnold Engineering in Tullahoma, Tennessee."

Chapter 10

1. *Skywriter* (North American Rockwell), October 22, 1965.
2. That was where I spent my time that week, so what I have written about the CARR came from my personal notes.
3. *Skywriter* (North American Rockwell), October 22, 1965, p. 1.
4. *Ibid.*, p. 3. I have paraphrased the information on AFRM 009 from an article on the front page and the captions under the nine photos.
5. *Skywriter* (North American Rockwell), October 29, 1965.
6. *Skywriter* (North American Rockwell), November 12, 1965.
7. *Ibid.*, p. 1.
8. *Skywriter* (North American Rockwell), November 19, 1965, p. 1.
9. *Ibid.* The Mag Robinson appointment was posted on the front page with an article that included details about Mag as well as his photograph.
10. "Apollo Spacecraft 009 Returns to North American after Successful Mission," *Skyline* 24, no. 1 (1966); *Skywriter* (North American Rockwell), February 26, 1966.
11. *Skywriter* (North American Rockwell), March 18, 1966.
12. What I have written here came from facts that I read in the March 29, 1968, *Skywriter* and from my engineering notes from the time that I spent designing the Mission Control Programmer that flew AFRMs 011, 017 and 020. The story of AFRM 020 can be found online by Googling "Apollo Spacecraft 020," "Apollo 6," or "Apollo mission AS-502."
13. *Skywriter* (North American Rockwell), February 11, 1966. A few words about the AFRM 009 mission will be added because, besides the test objectives of the Apollo Command Module and the Service Module, there were test objectives for the launch boosters and for the Apollo Boost Guidance. I cannot comment on any of the boosters or the guidance test results because they were not part of my team's responsibility. However, if any reader is interested in boosters and guidance information, you may find details online under "NASA Apollo Mission AS-201." For guidance, try "MIT Draper Labs guidance Apollo mission AS-201."

Chapter 15

1. NASA Electronics Research Center, "Apollo Command and Service Module Stabilization and Control System Design Survey: Design Criteria Program Stability, Guidance, and Control," December 20, 1968, pp. 1–21.
2. *Skywriter* (North American Rockwell), January 27, 1967.
3. *Skywriter* (North American Rockwell), April 21, 1967.
4. *Skywriter* (North American Rockwell), April 28, 1967.
5. John Uri, "55 Years Ago: The Apollo 1 Fire and Its Aftermath," NASA History, https://www.nasa.gov/feature/55-years-ago-the-apollo-1-fire-and-its-aftermath (updated February 4, 2022).
6. *Skywriter* (North American Rockwell), May 5, 1967.
7. *Skyline* 25, no. 4 (1967); "North American, Rockwell-Standard Plan Merger," *Chemistry English News* 45, no. 15 (1967): 21–22.

Chapter 16

1. The names of the members of Carl Conrad's unit came from the organization chart issued on June 7, 1968. I have in my possession fifteen sheets of the organization charts that cover August 1, 1965, to June 7, 1968.
2. "Progress in Downey," *Skywriter* (North American Rockwell), August 25, 1967.
3. The history, organization responsibilities, and personal notes that I have provided came from my association with the Apollo Program, working with many Apollo personnel: my engineering notes, my copies of the Guidance and Control organization charts and my recollection of events and colleagues. Remember also that this is a behind-the-scenes story that has never been told, so most of the names in my story may not appear in any publication.

Chapter 17

1. *Skywriter* (North American Rockwell), November 3, 1967; *Skywriter* (North American Rockwell), November 10, 1967; *Skywriter* (North American Rockwell), November 17, 1967. These editions have details on the launch, mission, Earth reentry, recovery at sea and Dr. Mueller's comments on the success of the mission.
2. Ed Kelley, North American Aviation Inc. letter 67MA6514, September 19, 1967.

Chapter 18

1. *Skywriter* (North American Rockwell), March 1, 1968, p. 3.
2. *Skywriter* (North American Rockwell), March 8, 1968.
3. *Skywriter* (North American Rockwell), March 15, 1968.
4. *Skywriter* (North American Rockwell), March 22, 1968.

Chapter 19

1. *Skywriter* (North American Rockwell), March 22, 1968.
2. Two editions of the company newspaper had details of the mission accompanied by photographs: *Skywriter* (North American Rockwell), April 5, 1968; *Skywriter* (North American Rockwell), April 12, 1968. See also "Apollo-Saturn Uncrewed Missions" at https://www.nasa.gov/mission_pages/apollo/missions/Apollo-Saturn-Uncrewed.html.
3. *Skywriter* (North American Rockwell), May 3, 1968.
4. *Skywriter* (North American Rockwell), May 31, 1968.
5. *Skywriter* (North American Rockwell), June 7, 1968.

Chapter 20

1. Ernst Stuhlinger and Frederick I. Ordway III, *Wernher von Braun: Crusader for Space* (Malabar, FL: Krieger, 1996), 231. During the years of the Surveyor Program, there were several publications of mission results, but I kept none of them. If readers Google "Surveyor to the Moon," they will find some interesting details on Surveyor.
2. T.A. Heppenheimer, "The Space Shuttle Decision: Chapter 6: Economics and the Shuttle," National Space Society, https://space.nss.org/the-space-shuttle-decision-chapter-6/. My statement that at the time of the Apollo Program it cost $1,000 to put a pound into orbit and that the plan was to reduce it to $100 a pound is based on the fact that those working on the Apollo Program (as well as the Space Shuttle Program) kept hearing that goal all the time.
3. I receive the *USC Trojan Family Magazine* and the *USC Viterbi* magazine, and in the 2004 magazines it was announced that Andrew Viterbi and his wife had made a $52 million donation to the USC School of Engineering, requesting only that the school be named the Viterbi School of Engineering. Andrew Viterbi founded his company Qualcomm Incorporated for the purpose of manufacturing the Viterbi Algorithm Microchip, which is sold worldwide for use in every cell phone, TV and many more items that use digital communications. On page 5 of my copy of the winter 2014 *USC Trojan Family Magazine* is an article announcing the tenth anniversary of the Viterbi donation with a photo of Andrew Viterbi and his wife Erna with USC president C.L. Max Nikias and his wife Niki. In the autumn 2017 issue of the *USC Trojan Family Magazine*, there is a full-page picture of Andrew Viterbi on page 49 with a cell phone to his ear. The caption on the top right-hand side of his face reads, "A Golden Anniversary." It continues: "There are 12.4 billion cell phone calls made every day on Earth. All of them share one unique property: the voices are decoded and made clear across great distances through the Viterbi algorithm. It's named after its inventor, Andrew Viterbi PhD '62, namesake of the USC Viterbi School of Engineering. The algorithm turned 50 this year, and it's still influencing modern communications. Learn how at bit.ly/ViterbiAlgorithm."
4. *Skywriter* (North American Rockwell), May 31, 1968.
5. *Ibid.*
6. *Skywriter* (North American Rockwell), June 14, 1968; *Skywriter* (North American Rockwell), July 12, 1968; *Skywriter* (North American Rockwell), August 2, 1968; *Skywriter* (North American Rockwell), August 16, 1968; *Skywriter* (North American Rockwell), August 23, 1968.

Chapter 21

1. The Apollo 8 description of the Customer Acceptance Readiness Review (CARR) I have provided comes from my memory of having attended a previous CARR for AFRM 009, but the date of the CARR was reported in a photograph with a caption in the June 14, 1968, issue of *Skywriter*.
2. *Skywriter* (North American Rockwell), August 16, 1968.
3. *Skywriter* (North American Rockwell), December 13, 1968.
4. *Skywriter* (North American Rockwell), January 3, 1969.
5. The January 3, 1969, *Skywriter* was almost exclusively devoted to Apollo 8, with many articles and two pages of photographs, which is where I obtained most of my information about this mission, augmented by details I got by Googling "Apollo 8." The reader who wants to know more about Apollo 8 can check it out online.
6. "Apollo Guidance and Navigation Field Training Aid," *AC Electronics* (1969); Nancy Atkinson, "The Story of the Apollo Guidance Computer, Part 3," Universe Today, August 9, 2019, https://www.universetoday.com/143113/the-story-of-the-apollo-guidance-computer-part-3/#:~:text=Lab%20engineer%20Ramon%20Alonso%20came,represented%20programs%2C%20verbs%20and%20nouns. Better yet, if the reader wants a good description of how the DSKY was used during a moon mission, I recommend Gene Farmer and Dora Jane Hamblin's *First on the Moon: A Voyage with Neil Armstrong, Michael Collins, Edwin E. Aldrin, Jr.*, in which Michael Collins

(who was the capsule pilot and navigator who operated the DSKY) describes how he made his entries into the keyboard when the capsule was on its way to the moon.

7. This account is based on the January 3, 1969, *Skywriter* (verified online), as well as my engineering notes and my recollection of having worked on Apollo 8.

8. Quoted directly from *Skywriter* (North American Rockwell), January 3, 1969, p. 4.

9. I recall vividly the prayer that Frank Borman read from space during Christmas because of my association with Apollo 8, so I just went to a Bible and retrieved it. See also David R. Williams, "The Apollo 8 Christmas Eve Broadcast," NASA Goddard Space Flight Center, https://nssdc.gsfc.nasa.gov/planetary/lunar/apollo8_xmas.html (updated September 25, 2007).

10. *Skywriter* (North American Rockwell), January 3, 1969.

11. *Skywriter* (North American Rockwell), January 10, 1969.

12. *Skywriter* (North American Rockwell), February 21, 1969; *Skywriter* (North American Rockwell), February 28, 1969.

13. *Skywriter* (North American Rockwell), March 14, 1969.

14. *Skywriter* (North American Rockwell), March 14, 1969; *Skywriter* (North American Rockwell), March 21, 1969; *Skywriter* (North American Rockwell), March 28, 1969; *Skywriter* (North American Rockwell), April 4, 1969.

15. *Skywriter* (North American Rockwell), June 6, 1969; *Skywriter* (North American Rockwell), June 13, 1969.

16. *Skywriter* (North American Rockwell), May 16, 1969; *Skywriter* (North American Rockwell), May 23, 1969.

17. *Skyline* 24, no. 1 (1966).

Chapter 22

1. The use of rope memory and the size of the memory in the Apollo Command Module Computer quoted here and elsewhere can be found online under "Ramon Alonso" or "Apollo CMC." See also Charles Fishman, "The Guts of NASA's Pioneering Apollo Computer Were Handwoven Like a Quilt," Fast Company, June 14, 2019, https://www.fastcompany.com/90363966/the-guts-of-nasas-pioneering-apollo-computer-was-handwoven-like-a-quilt.

Chapter 23

1. Unit Organization Chart (1969), author's files.

2. *Skywriter* (North American Rockwell), January 24, 1969, pp. 1, 3.

3. *Skywriter* (North American Rockwell), March 21, 1969; *Skywriter* (North American Rockwell), April 25, 1969.

4. *North American Rockwell News*, July 11, 1969.

5. *North American Rockwell News*, July 18, 1969; *North American Rockwell News*, July 29, 1969.

6. *North American Rockwell News*, July 11, 1969. The reader can Google each astronaut's name for an online personal history.

7. "Florida Problem: Apollo Crowds," *North American Rockwell News*, July 11, 1969.

8. Gene Farmer and Dora Jane Hamblin, *First on the Moon: A Voyage with Neil Armstrong, Michael Collins, Edwin E. Aldrin, Jr.* (Boston: Little, Brown, 1970). If the reader wants to learn the true story of Apollo 11, I recommend this book.

9. "Space Division Team Will Attend Module to Make It Safe to Handle," *North American Rockwell News*, July 18, 1969.

10. The reader may recall from chapter 12 that, during my work on the unmanned Apollo capsules, I was offered a position as a technical specialist from a manager at Aerojet General Corporation in Azusa, California. At that time, I researched Aerojet and found out that the company was founded by Dr. Theodore von Kármán and that he was a famous engineer. I also read about his saying on the difference between scientists and engineers. In addition, Dr. von Kármán is mentioned at least three times in the book *Wernher von Braun: Crusader for Space* (referenced previously). The reader might want to learn more about Dr. von Kármán by Googling his name—a very interesting person.

Bibliography

Boole, George. *An Investigation of the Laws of Thought.* New York: Barnes & Noble Books, 2005. (Originally published in 1854.)

Campbell, John C., and Douglas M. Higgins, eds. *Mathematics.* 3 vols. Belmont, CA: Wadsworth, 1984.

Collins, Michael. *Liftoff: The Story of America's Adventure in Space.* New York: Grove Press, 1988.

Division General Precision, Inc. *Technical Information for the Engineer: Servo Motors, Tachometer Generators, Motor Generators, Synchros.* Little Falls, NJ: Kearfott, 1959 (Sixth edition 1960).

Dreyer, J.L.E. *A History of Astronomy from Thales to Kepler.* Second edition. New York: Dover, 1953.

Farmer, Gene, and Dora Jane Hamblin. *First on the Moon: A Voyage with Neil Armstrong, Michael Collins, Edwin E. Aldrin, Jr.* Boston: Little, Brown, 1970.

Glenn, John, with Nick Taylor. *John Glenn: A Memoir.* New York: Bantam Books, 1999.

Guillen, Michael. *Five Equations That Changed the World.* New York: MJF Books, 1995.

Kraft, George D., and Wing N. Toy. *Microprogrammed Control and Reliable Design of Small Computers.* Englewood Cliffs, NJ: Prentice-Hall, 1981.

Mano, M. Morris. *Computer System Architecture.* Second edition. Englewood Cliffs, NJ: Prentice-Hall, 1982.

Mazda, F.F., ed. *Electronics Engineer's Reference Book.* Fifth edition. London: Butterworth & Co., 1983.

National Geographic History. "Aztec Age: Rise of an Empire." July/August 2021.

Nelson, E.W., C.L. Best, and W.G. McLean. *Schaum's Outline of Theory and Problems of Engineering Mechanics.* New York: Schaum Publishing Company, 1952.

Schaum, Daniel, and Carel W. Van der Merwe. *Schaum's Outline of Theory and Problems for Students of College Physics.* New York: Schaum Publishing Company, 1936. (Fifth edition 1946.)

Shepard, Alan, and Deke Slayton, with Jay Barbree and Howard Benedict. *Moon Shot.* New York: Turner Publishing, 1994.

Stuart, Gene S. *The Mighty Aztecs.* Washington, DC: National Geographic Society, 1981.

Stuhlinger, Ernst, and Frederick I. Ordway III. *Wernher von Braun: Crusader for Space.* Malabar, FL: Krieger, 1994. (Reissued in 1996 with corrections.)

Thomas, George B., Jr. *Calculus and Analytic Geometry.* Second edition. Reading, MA: Addison-Wesley, 1953.

Wu, Margaret S. *Introduction to Computer Data Processing.* New York: Harcourt Brace Jovanovich, 1975.

Index